THE ROOTS OF MODERN ENVIRONMENTALISM

THE CROOM HELM NATURAL ENVIRONMENT —
Problems and Management Series

Edited by *Chris Park, Department of Geography,
University of Lancaster*

THE ROOTS OF MODERN ENVIRONMENTALISM

DAVID PEPPER

With material contributed by
JOHN PERKINS and
MARTYN YOUNGS
Consultant Editor:
TOM COLVERSON

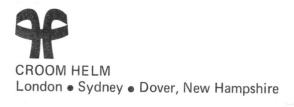

CROOM HELM
London • Sydney • Dover, New Hampshire

©1984 D. Pepper,
Croom Helm Ltd, Provident House, Burrell Row,
Beckenham, Kent BR3 1AT

Croom Helm Australia Pty Ltd, First Floor,
139 King Street, Sydney, NSW 2001, Australia

British Library Cataloguing in Publication Data

Pepper, David
    The roots of modern environmentalism.
    1. Environmental protection − History
    2. Conservation of natural resources
    − History
    I. Title   II. Colverson, Tom
    333.7'2   TD170

    ISBN 0-7099-2064-4

Croom Helm, 51 Washington Street, Dover,
New Hampshire 03820, USA

Library of Congress Cataloging in Publication Data

Pepper, David.
    The roots of modern enviromentalism.

    Bibliography: P.
    Includes index.
    1. Ecology − Philosophy − History. 2. Enviromental
Protection\− Philosophy − History. 3. Nature Conservation −
Philosophy − History. I. Perkins, John W. II. Youngs,
Martyn J. III. Title.
QH540.5.P47 1984   304.2'09   84-19867
ISBN 0-7099-2064-4

Printed and bound in Great Britain
by Billing & Sons Limited, Worcester.

CONTENTS

* Includes material contributed by John Perkins

+ Includes material contributed by Martyn Youngs

LIST OF FIGURES

To Lucy and Edward; and to all my 'Man-Environment Attitudes' students, past and present.

FOREWORD AND ACKNOWLEDGEMENTS

This book is designed to be of interest to students, in the *broadest* sense, of the environment and environmental matters. It is about historical, philosophical and ideological aspects of environmentalism. I have written it because I think that those who participate in and study current environmental issues and debates often lack a perspective which stems from adequate knowledge of these aspects. My justification for their study is laid out in the Preface - a part of the book which the very beginning student might do well to skim through initially and return to after he has read what follows. Because it is written for students rather than for academics established in the field of environmental attitudes, the book tries to render some complicated ideas into relatively straightforward prose. It will not, therefore, be entirely free of the pitfalls which commonly attend such an attempt. The degree of simplification which is a prerequisite for understanding can so easily become oversimplification; the repetition which helps in familiarisation with ideas can become tedious. I hope that neither of these failings are present to an unacceptable degree. If they are, this is very much *my* failing because my principal task here has been to draw together the ideas and perspectives of others and to synthesise them into a work which makes sense.

One of the 'others' whose perspective has informed the work is John Perkins. For some years he has excited the imagination and interest of my students of environmentalism, with his science historian's approach and ideas and the thoughtful and lucid way in which he has imparted them. His contribution to this book does not merely lie in the material which he has gathered for Chapter 2; he has also brought considered and constructive criticism to bear on a number of other chapters. Much the same applies to Martyn Youngs, who has not only contributed material to the work, but has also been a supportive and stimulating academic colleague.

While he has not been involved with the present text, Denis Cosgrove of Loughborough University has nonetheless had a pervasive influence on it. When he taught human geography at Oxford Polytechnic, Denis' imagination and scholarliness was primarily responsible for leading me - a physical geographer and rather unimaginative environmentalist -

towards a realisation of the importance of the social, material and ideological aspects of environmental 'problems'. As a result, the 'Man-Environment Attitudes' course which we developed, and on which this book is based, acquired a distinctive character and touched the lives of many of its students. I thank John, Martyn and Denis, and hope that my written work has done justice to their material and their influences.

On the editorial side, Tom Colverson has contributed expert advice and judgement, and unflagging hard work; Joy Dalton has operated Lasercomp with admirable finesse and has given skilful and alert editorial counsel (she has also shown an almost uncanny, and quite unfair, ability to spell accurately). Finally, I am greatly indebted to Heather Jones for drawing the figures, and to Alan Jenkins and John Gold for their consistent support and help during the writing of this book.

I acknowledge gratefully the following, for permission to reproduce or redraw their copyright material: Figure 1, *The Journal of Geography*; Figures 2 and 8, *The Journal of Geography in Higher Education*; Figure 3, John Wiley and Sons, Inc.; part of Figure 5, Pergamon Press Ltd.; Figure 7, IBM United Kingdom Ltd.

David Pepper, Oxford Polytechnic 1984

ON WHY HISTORY AND PHILOSOPHY MATTER

1. THE 'FACTS' OF THE MATTER

'The heart of the countryside - ripped out', said a leaflet produced by Friends of the Earth in 1983. 'For centuries we lived in harmony with our countryside'...but 'In just one generation everything has changed. Destruction of habitats has rapidly escalated to a disastrous level. The effect on wildlife and countryside has been devastating'. There followed a list of facts and figures on how many acres of farmland had been lost in Britain, and how many wildlife species had been destroyed by modern agribusiness techniques.

Over the past 15 or so years countless volumes of similar facts and figures have been produced in the West, documenting in detail many facets of a perceived post-war deterioration in man's physical environment. This deterioration has been attributed to widespread industrialisation over the past hundred or two years, allied to increased material consumption by rapidly-expanding populations. Concern has been expressed not only by environmental groups like Friends of the Earth, but also by industrialists - like those of the Club of Rome which sponsored the notorious *Limits to Growth* report in 1972 - and government agencies.

But after data have been collected, and concern has been voiced, the central question for environmentalists has been 'So what? What is likely to be *done* to remedy environmental problems like the "destruction" of the countryside?' One school of thought is represented by Marion Shoard, who believes that if only people *know* about what is going on then remedial action will naturally follow:

> there is no public support for the agricultural 'reclamation' which is destroying the countryside, and there never will be, however vociferous the farming lobby becomes. It is only because the people do not know what is going on that the process continues (Shoard 1980).

Lord Melchett, however, is less sanguine. He says:

> Conservationists need to realise that scientific evidence, rational arguments and compromise do not win political arguments (Melchett 1981).

In other words, sheer volumes of data - of facts and figures - are, says Melchett, unlikely to be very persuasive in themselves. For, as the contemporary environmental debate has shown, people have an almost infinite capacity either to ignore or to heed selectively the 'facts of the matter'. Having *first* made up our minds, frequently from an irrational base, about what we want, we all tend to look for 'facts and figures' to support our position, from which we will be dislodged only very gradually if at all. Thus, we argue, and perceive the arguments of others, not in an objective and unbiassed way. We have presuppositions, or even vested-interest positions that colour our perception of the facts. These are frequently economically based, as in the case of the pro-agribusiness lobby which has such an effective anti-conservation voice in the corridors of power, but they are also shaped by a host of non-economic factors. These will be socially and culturally derived, and imparted to us via our education and socialisation. Anyone who wants to influence us or change our minds will have to understand and take account of these presuppositions and vested interests. It will be no good bombarding us with 'facts' which we are anyway predisposed to dismiss. A wiser strategy would be to shake the foundations of our beliefs by undermining the assumptions on which they are based. And this is how we should approach differences with other people. As Bertrand Russell put it:

> When an intelligent man expresses a view which seems to us obviously absurd, we should not attempt to prove that it is somehow not true but we should try to understand how it ever came to *seem* true. This exercise of historical and psychological imagination at once enlarges the scope of our thinking, and helps us to realise how foolish many of our own cherished prejudices will seem to an age which has a different temper of mind (Russell 1946, Chapter 4)

In other words, we should listen to what others say, and reflect not necessarily upon the 'truth' of their arguments, but on *why* they make them and believe in them, i.e. from what material or ideological vested interest position they speak, and what broader assumptions and philosophy serve this interest. And if we wish to influence their thinking, we shall have to study the history of how their thinking came to be as it is - for we cannot effect a process of change without first knowing how changes came about in the past. Harvey puts it differently. He says:

> Concepts and categories cannot be viewed as having an independent existence, as being universal abstractions true for all time. The structure of knowledge can be transformed, it is true, by its own internal laws of transformation (including the social pressures internal to science). But the results of this process have to be interpreted in terms of the relationships they express within the totality of which they form a part... It is irrelevant to ask whether concepts, categories and relationships are 'true' or 'false'. We have to ask, rather, what it is that produces them and what is it that they serve to produce? (Harvey 1973 p 298)

The implication of this is that if we seek for the future the kind of *real* social and environmental changes which much of the standard environmentalist literature calls for, and which have to be accompanied by widespread attitudinal changes, then we must develop an *historical* perspective on how we and others have arrived at our present set of attitudes, and understand what material changes will be needed to help foster a new set. We are, then, hearing an argument for the practical usefulness and social relevance of historical and philosophical study. Not surprisingly, the historian Keith Thomas also argues in this way. 'Concern for the natural environment', he says, is normally regarded as relatively recent:

> Nowadays one cannot open a newspaper without encountering some impassioned debate about culling grey seals or cutting down trees in Hampton Court, or saving an endangered species of wild animal. But to understand these present-day sensibilities we must go back to the early modern period. For it was between 1500 and 1800 that there occurred a whole cluster of changes in the way in which men and women, at all social levels, perceived and classified the natural world about them... It was these centuries which generated both an intense interest in the natural world and those doubts and anxieties about man's relationship to it which we have inherited in magnified form (Thomas 1983, p 15)

This book accepts all these arguments. Therefore it is not about the 'facts' of our environmental predicament in the West. Instead, it accepts that a study of the history and philosophy of environmental *ideas* provides an invaluable perspective to those who are attempting to find a way out of our predicament. On the other hand, a study of the facts alone seems to lead nowhere. Scores of books have been written about these facts - of chronic imbalances in population/resource ratios, of ecologically damaging technology, of wasteful consumption patterns - such that substantial acreages of forest must have been consumed in the process. Yet one can legitimately argue that little change of a truly *fundamental* nature has been achieved by the environmental movement (and we include in this movement those who have attempted to reverse the arms race). The spread of detailed knowledge about how man degrades and threatens his own planet has not of itself produced the likelihood of serious or permanent remedial action. For this reason we now argue that environmentalists should be wary and weary of the details contained in yet more 'gloom and doom' publications such as the *Global 2000 Report to the President* (US Interagency Committee 1982). Its 750 pages of facts about resource depletion and population growth are all subordinate to the one piece of information which constitutes the publisher's eyecatcher on the dustjacket, and simply says 'Commissioned by Carter, disregarded by Reagan'.

Not that we should dismiss this kind of report out of hand, as Julian Simon (1981) does. But it will be assumed that readers of this book are already largely familiar with the standard environmentalist literature (like *Limits to Growth, Blueprint for Survival* and *Small is Beautiful*), and require only a brief review of the salient points (Chapter 1) before going on to examine, in Chapters 2, 3 and 4, the roots of some of the ideas in this literature. In other words we want to introduce those who know little of history and philosophy to a historical and philosophical perspective on environmentalism, for it is time that more environmentalists enlarged the scope of their thinking by exercising what Russell called the 'historical and psychological imagination' in respect of ideas which they so frequently advocate.

2. THE IMPORTANCE OF IDEAS AND WHERE THEY COME FROM

In a controversial article on the 'Historical Roots of our Ecologic Crisis' (the existence of such a crisis was assumed) Lynn White (1967) wrote 'What we do about ecology depends upon our *ideas* of the man-nature relationship' (emphasis added). This statement asserts the practical importance of the study of the history and philosophy behind our conceptions of the man-nature relationship in averting the perceived crisis.

Other environmentalists have taken up White's cue. They have pointed out that the kind of society which many see as ecologically feasible and stable, and therefore socially desirable, would be based on ideas and values which may already have existed for a very long time, but nonetheless run counter to those that form the basis of the prevailing economic ethos and conventional wisdom. Among these are the notions of no-growth economies and populations, decentralisation and small-scale communities, the blurring of the town-country distinction, alternative/intermediate/appropriate and environmentally sympathetic technology, recycling of wastes and raw materials, cooperative life styles and - most important - the idea that individual advancement and social progress are measureable not so much in material terms, but are attainable through social justice and individual mental and spiritual fulfilment. As Lowe and Goyder (1983 p25) say, 'In expressing concern about pollution, the destruction of nature, the loss of amenity and the depletion of resources, environmentalism either explicitly or implicitly challenges existing assumptions about progress which equate material prosperity with general wellbeing'. Indeed, this challenge amounts to 'an attack on the central values and beliefs of industrial capitalism' (Cotgrove and Duff 1980). Instead of materialism, there is 'quality of life', including heightened environmental quality, and this is to become the substitute for the impending deterioration in material quantity in our lives which is seen as inevitable and desirable.

Most people have not rushed to embrace these ideas, which were enunciated with such stirring clarion calls in the early 1970s. They cling instead to notions of, and pine for the days of, economic growth. High technology progresses (in 1980-81, UK government grants for renewable energy development dropped to £11m, while £170m was available to spend on nuclear research (Hales 1982)). Centralisation, the division of labour, materialism, competition, and all the other ideas which some environmentalists deprecate seem as strongly-entrenched as ever despite mounting evidence that they do not 'work' for the majority of people. Schumacher (1973) distilled all the 'anti-environmental' values into six main ideas, which he described as 'all stemming from the nineteenth century, [and] which still dominate, as far as I can see, the minds of "educated" people today'. These were evolution; competition, natural selection and the survival of the fittest; the Marxist belief in the materialist base of history; the Freudian emphasis on the overriding importance of the subconscious mind; the ideas of relativism, 'denying all absolutes, dissolving all norms and standards'; and the philosophy that 'valid knowledge can be attained only through the methods of the natural sciences'. Capra (1982) distilled such ideas even further, believing that their essence lay in the 'Cartesian-Newtonian Paradigm' in science (see Chapter 2) which has dominated the way of thinking of most people, scientists and non-scientists, in the 20th century - even though a growing band of scientists now recognise the inadequacies of this paradigm (a glossary of terms such as 'paradigm' will be found in the appendix to this book).

Schumacher and Capra represent *par excellence* that tradition in environmentalist literature in which essentially practical men recognise and write of the prime importance of ideas and values in determining environmental actions and policy in the West. In view of this, it is perhaps not surprising that both of these *idealists* conclude that the remedy for our environmental predicament lies in a change of values - in what *ideas* we hold as true and valid and worth defending. Yet this remedy seems a little lame, for it is offered without accompanying practical guidance on *how* such a change is to be achieved. Perhaps it was a frustration with such lame conclusions - rife in the 1970s environmentalist literature - that led Henryk Skolimowski (actually a man of ideas - a philosopher) to comment, 'Witness the endless, insipid, impotent discussions about values, in which pretentious and sentimental claims are made to the effect that if only we changed our hearts a little and became more charitable to each other, all will be well' (Skolimowski 1981 p59). He was characterising this type of approach as superficial, in that it does not consider adequately what is the source of our ideas, and if we do not know how our present values came about in the first place then we are unlikely to be able to change them. With his background, Skolimowski focusses attention on our 'cosmology' - our view of the world and the universe and how we relate to them - as being the source of our values. Our cosomology is a function of our underlying philosophy, and this is one which, he says, is at present far too heavily based on empiricism and scientism and is too mechanistic and analytic (see Glossary and Chapter 2). It is too little based on 'humanistic' notions of morality towards nature and it is one which shuns a discussion of values, preferring to give primacy to 'factual' knowledge. Like White and Thomas, therefore, Skolimowski thinks it important to try to trace historically the sources of our present philosophies - our cosmology. He traces the eclipse of values by 'facts' back to the 19th century, attaching it (like White, rather vaguely) to the growth from the 17th century of the science and technology of the type we know today.

There are others, however, who in their turn would regard this approach as too superficial. For one thing it seems to regard values as somehow a function of technology For another, although it does point out that 'Science did not develop in a social vacuum but as part of the unfolding new culture' (Skolimowski 1981 p8), it does not go on to discuss what material aspects of that culture lent support to *particular* ideas and philosophies (now identified as ecologically harmful) as against *others* that existed alongside them but were neglected (and are now seen as ecologically benign).

Some writers see this material base as the critical factor in influencing how we got our present ideas and values towards nature, and consequently how to change them. They see our attitudes as crucially influenced by our economic system (of which technology is but one manifestation) - i.e. the way we organise and obtain our material existence. And they go on to assert that it is this which creates specific and particular relationships between people and between people and nature. It is for this reason that Alan Schnaiberg's book on *The Environment - From Surplus to Scarcity* (1980) focusses its attention on the 'treadmill of production' - a term derived from the economist J.K. Galbraith's view of how our present materialist and consumer-oriented society operates. Schnaiberg sees that the material methods and aims of this society have a profound influence on how it perceives its members, and nature. This kind of *materialist* perspective is an important development upon the idealistic views of writers like Schumacher and Capra, and it will be examined in Chapter 6.

3. THE CULTURAL FILTER AND THE ROLE OF SCIENCE IN IT

The discussion so far suggests that it is of prime importance for us to study, as well as the 'real' or tangible physical environment, how different groups and individuals *perceive* that environment, and the nature of the economically- socially- and culturally-based presuppositions which colour this perception, or as some express it, the *cultural filter*. To use this term is to draw on a concept which is very important in social sciences and history.

This concept suggests that in taking an historical perspective on the development of ideas and values it is important not to see people of earlier times as thinking in essentially *the same way* as ourselves, but merely transposed to another age. Instead we must, as Thomas (1983 p16) puts it, attempt 'to reconstruct an earlier mental world in its own right' and to expose the assumptions, some barely articulated, which underlie the perceptions, reasonings and feelings of earlier peoples.

When we can do this for earlier peoples, we may find it easier to do it for ourselves, rather than assuming that our own world view is the only possible, natural and right one for all time. For we, too, have a series of what Lovejoy calls 'implicit or imcompletely explicit assumptions, or more or less *unconscious mental habits*, operating in the thoughts of an individual or a generation' and these constitute *our* cultural filter. The filter is of prime importance, for:

It is the beliefs which are so much a matter of course that they are rather tacitly presupposed than formally expressed and argued for, the ways of thinking which seem so natural and inevitable that they are not scrutinised with the eye of logical self consciousness, that often are the most decisive of the character of a philosopher's doctrine, and still oftener of the dominant intellectual tendencies of an age (Lovejoy 1974 p7).

Because this concept involves our thinking consciously about, and questioning, 'ways of thinking that seem so natural and inevitable', it is an extremely difficult concept to grasp 'from cold'. But one of the purposes of this book is to exemplify the concept, to make it more familiar. And because, as Lovejoy (p18) says, 'These endemic assumptions, these intellectual habits, are often of so general and so vague a sort that it is possible for them to influence the course of man's reflections on almost any subject', it is essential to discover what they are. For the question of how we perceive and ought to act to change our environment and our landscapes is not free from these endemic assumptions. As Pryce (1977) puts it, 'From time immemorial people have formed opinions, developed attitudes and based their actions on images that may have borne little or no resemblance to reality. Biassed or, indeed, completely erroneous ideas concerning the... environment are potentially just as influential as those conforming to the real world'. That is, as White said, what we do about nature and ecology is a function not of what *is* 'out there', but of how we perceive it, and our perception is a function of our cultural filter and the assumptions in it. Thus:

> Man consciously responds to his environment as he perceives it: the perceived environment will usually contain some but not all of the relevant parts of the real environment, and may well contain elements imagined by man and not present in the real environment ... The real environment ... is seen through a cultural filter, made up of attitudes, limits set by observation techniques, and past experience. By studying the filter and reconstructing the perceived environment the observer is able to explain particular options and actions on the part of the group being studied. (Jeans 1974, see Figure 1).

The last sentence is an eloquent justification for the approach taken in this book, and the definition of the cultural filter makes it clear that the philosophies and history of the group being studied, and its resultant 'cosmology' and ideologies assume prime importance in the filter, and therefore in man-environment studies. As Mills (1982) says about earlier perceptions of nature, it is 'no more a book or a giant human being than it is an extraordinarily complex machine. That certain societies should find such views of it convincing, however, is highly informative, and provides us with a direct means of knowing their central needs and aspirations'.

But then, to consider the notion of a perceived and a 'real' environment raises a host of further questions, which will come into the following chapters, especially Chapter 5. For example, if there is a difference between real and perceived environments - real and perceived nature - and if the latter is 'potentially just as influential' as the former in determining what we do about, and to, our environment, then this soon begs the question of just how important is 'reality' at all, and how can we in fact tell the *difference* between what is 'perceived' and what is 'real'. This leads on to make us ask what 'reality' is, and what 'facts' are, and how we can possibly know, objectively, 'facts' about our 'real' environment. We enter, right away, a fundamental philosophical debate with practical implications which echoes the views of Kant (1724-1804) that we cannot know the world as it is (the 'noumenal' world). We can know only the 'phenomenal' world - that one presented to us through our senses and mediated by experience.

A prime channel (or 'epistemology' i.e. an approach to knowledge) by which we inform ourselves about our 'real' environment, animate and inanimate, is constituted by *science*. Science furnishes us with both a method to study the environment, and a philosophy which tells us about our relation to the environment. We are used to thinking that this method is the means which gives us objectively facts about our real environment, and that, as a philosophy, science puts great value on objective knowledge. But is science 'objective'? If something has been discovered or determined scientifically is it then 'true'? Is it part of our 'real' environment? We will learn more of the answer to these questions in Chapters 2 and 5, but without thinking much about it we can say that in Western cultures we incline to say 'yes' to them all. Furthermore we usually feel that scientific knowledge is - by virtue of its 'objectivity' - highly to be valued. If something is 'scientifically proved', then it becomes in our eyes valid and legitimate - as the man wearing a white coat in the toothpaste advert will tell us, while he reels off an unintelligible 'formula' (which by virtue of its very unintelligibility renders itself, in our eyes, all the more inaccessible, impressive and authoritative, and therefore the province of the scientific 'expert' rather than of the ordinary mortal). Conversely, if we wish to denigrate the arguments of our opponents, we have only to dub them 'unscientific'. And if we want to completely discredit them we need but to charge them with one of the most heinous crimes of our age; that is, of being 'subjective' or 'emotional' - i.e. not value-free.

So in Western eyes, science is the repository of truth. It is often the monolithic arbiter of what is and is not valid. The scientific 'expert' is its high priest, and it is he who can tell us objectively what is the difference between our 'real' and our perceived environment. This popular notion derives ultimately from Francis Bacon's view of the properties and characteristics of science and scientists. But it is a view which is coming very much under critical scrutiny, on several grounds and from several sources.

Indeed, those who are interested in the fate of the earth should be instantly suspicious of it. For on the one hand 'science' in the hands of the scientists and technologists of, for example, the big energy corporations, tells us that man's relationship to nature is one of *domination and exploitation* - and can continue increasingly to be so, given that we perfect to even greater heights our technology and our environmental 'management' techniques. But on the other hand, 'science' in the hands of the ecologists and environmentalists tells us that we are essentially part of nature and in a *reciprocal* relationship to it.

'Real' and perceived environments differ. The latter is the important influence on decision making. Environmental perception is different in different cultures – humans perceive nature through their cultural filter.

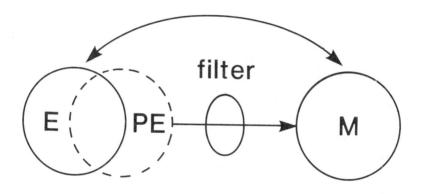

E   Environment
PE   Perceived Environment
M   Man

FIGURE 1   **The Cultural Filter (source: Jeans 1974)**

If we dominate rather than become an equal partner with it - if we act as 'despots' rather than as 'stewards' (Passmore 1974) - then we will be in trouble because the damage which we do to the system will reverberate on us due to the fact that we are in and not apart from the system.

How can this 'science' which we have grown to respect for its precision, its unambiguity and its neutrality give us these two quite opposing messages at the same time? If science is 'true', how can both messages be simultaneously true? There is now a wealth of literature to show us that science and scientists are *not* impartial, value-free or objective. At one level, critics of the classical science which we learn in school will say that it is not value-free because, as a philosophy, it shuns debates about values - and this represents a value in itself (i.e. that values are not as important as 'objective facts') (Harvey 1974A). Thus a scientist could happily work to develop nuclear technology, or simply to study the way it works, but he might well not concern himself as a professional with how and for what purposes this technology is to be used. Although this 'Eichmann mentality', as Harvey calls it, has been displayed in the past, there have also been growing movements for social responsibility to be displayed in scientific work, which have aimed to get scientists involved professionally with the values behind the uses to which their work has been put (BSSRS 1982).

Related to this type of criticism of scientific objectivity has been the observation that while the work of scientists to answer questions and solve problems may proceed in a neutral way, the issue of *what* questions are asked, and what problems are deemed to be worthy of scientific study is, of course, not devoid of value judgements. Johnston (1979) raises it in respect of dominant research paradigms in geography, and it can be raised for every academic discipline with or without scientific pretensions.

Such aspects of the non-neutrality of science may be distilled largely into sins of omission. However, this book will also be concerned (e.g. in Chapter 7) with criticisms that people have, in a more active and environmentally destructive way, *used* science, with its images of respectability and truth, to hide or to justify and legitimate what may be very partial, unscientific and ideologically-derived policies. Thus the British Conservative Government in 1980-1 put so much research money into nuclear power rather than 'soft' energy sources perhaps because of the power of the pro-nuclear lobby, and also because it wanted to break the political power of the coalminers. But of course its spokesmen did not put this forward as the *justification* for its policy. Rather, they attempted to argue that in a 'scientific' (and this includes economic as well as technological) evaluation nuclear power had big advantages over other sources.

4. THE APPROACH AND RATIONALE OF THIS BOOK

To present a metaphor of the arguments put above, they say that much time has been spent by environmentalists in examining and presenting a *picture* of nature and man's influence on it as seen through a sort of 'mental instrument', like a microscope or telescope. However, they should now spend more effort in examining not the picture, but the characteristics of the 'microscope or telescope' *by which* they and others obtain the picture. This 'instrument' is the cultural filter - a world view - which is composed of economically, socially and culturally conditioned sets of assumptions and presuppositions which are rarely overtly examined and challenged. By looking at the history of the development of commonly-used cultural filters in environmental debates, we should be able to understand more clearly what we need to do to achieve the shift in attitudes which it is generally agreed is necessary to attain a more socially and ecologically harmonious society. This study, then, is not a rarified academic irrelevance; it is a practical and socially useful exercise in self-education.

It is necessary, too, to combat what is the most pernicious educational doctrine of the present (nuclear) age - that is, that there is basically nothing which can be done to alter the world as it is. This idea, that our power to effect social and environmental change is strictly limited, will be constantly returned to. For it is a very suspect idea, though a major element of the 'hidden curriculum' in our education (see Chapter 8). It is fostered by an education system which feeds all sorts of assumptions into our 'cultural filter' but seldom encourages us to examine or question those assumptions. What we do not examine, we tend not to question, and what we do not question we are unlikely to believe is amenable to change. Henry David Thoreau, who has inspired the environmentalist movement, put it better:

> It appears as if men had deliberately chosen the common mode of living because they preferred it to any other. Yet they honestly think there is no choice left. But alert and healthy natures remember that the sun rose clear. It is never too late to give up our prejudices. No way of thinking or doing, however ancient, can be trusted with proof. What everybody echoes, or in silence passes by as true today may turn out to be false tomorrow, mere smoke of opinion ... What old people say that you cannot do, you try and find that you can. Old deeds for old people and new deeds for new. (Thoreau 1974 p18).

The vital environmental and social question of our time becomes, then, how can we develop 'an alert and healthy nature' so that we can develop and use new ideas? As Russell suggested, we need to exercise our *historical* and psychological imagination in order to enlarge the scope of our *present* thinking.

This book will attempt to do this by synthesising disparate material which is relevant to the history and philosophy of environmentalism, and by making links between the past and the present. It will employ different analytical frameworks (see Chapter 1) which have been found in undergraduate teaching to increase understanding of difficult material by non-specialists.

It begins, in Chapter 1, by introducing O'Riordan's framework for classifying environmentalists and their ideologies. This distinguishes between technologically optimistic environmentalists - 'technocentrics' - and those who base their less sanguine views on a mixture of ecology and the non-scientific philosophy of romanticism; they are the ecological environmentalists, or 'ecocentrics'. The emergence of technocentric thought is traced to the development of rationalism and the scientific revolution of the 16th century onwards (Chapter 2), and attention is focussed on attitudes of dominance over nature which stemmed from them. The contrasting ecocentric ideology, of equality between man and nature, or subordination of the former to the latter, is then traced to its romantic (Chapter 3) and scientific roots - the latter based particularly on Malthus and Darwin (Chapter 4).

But out of this attempt to understand environmentalism's roots comes the paradox referred to earlier, of how science gives us two opposing messages about the man-nature relationship. This can be resolved only by questioning the notion of science as a source of objective knowledge about nature (Chapter 5). We are also drawn to examine the idea of objectivity itself, and we see that it is connected with a fundamental philosophical distinction - between *determinist* philosophies that see man and nature as separate entities, and *free will* philosophies, which in their most developed form see no separation between man and nature. This question of how much man is a creature of free will has political ramifications for all those, like the more radical environmentalists, who seek to effect social change. To exercise our will, and bring about change, we need to have an idea of how and why change has occurred in the past, and the historical and material perspective of Marxism (Chapter 6) is invaluable in trying to assess how valid is the environmentalist's model of social change (most commonly this is a pluralist and idealist model). Social and political change are linked, hence Chapter 7 examines the political roots of the more radical environmentalists, in an attempt to see whether their ideas represent more traditionally left- or right-wing ideologies. Finally, environmentalists themselves frequently see education as the primary vehicle for change in ideas, and therefore by implication 'environmental education' is regarded as some kind of panacea. But in Chapter 8 we criticise the notion that education of itself is a force for change rather than a vehicle for imparting, as Schumacher says, the reactionary ideas of the 'hidden curriculum'.

In all this we shall be examining the history and development of ideas, particularly the ideas represented in scientific thought since these constitute such a dominant part of our cultural filter. In following this examination the reader should bear very much in mind Lovejoy's warnings (1974 p21-2) about the pitfalls of studying the history of ideas, and realise that they apply to this work. Because it aims at unification and interpretation 'and seeks to correlate things which often are not on the surface connected, it may easily degenerate into a species of merely imaginative historical generalisations'.

'The history of philosophy and all phases of man's reflection is, in great part, a history of confusions of ideas'. In other words, we rush in where angels fear to tread, intending not to make a totally authoritative statement, therefore, but rather to stimulate the reader into further study of this fascinating area.

CHAPTER 1

MODERN ENVIRONMENTALISM

1.1 DEFINITIONS

If we read much into the debate about 'the environment', we soon become aware of several factions at work. Each professes concern over the state of the 'natural' as well as the man-made environment, and agrees that our relationship with nature has, in some way, to be improved. But there the accord frequently ends and warfare breaks out. One has only to read Fry's (1975) scathing remarks about Marxists or Bookchin's (1980) vituperative comments about middle-class environmentalists - or even the popular American car sticker which says 'If you're hungry and out of work, go eat an environmentalist' - to see that there is often no love lost between human beings who may have in common that they all think we should behave with more peaceful and harmonious intentions towards nature. Frequently this disagreement is a function of genuine ideological differences, but sometimes it simply results from ignorance or ill-based assumptions about what the other side stands for. This is partly because there is no clear-cut and easily circumscribed definition of environmentalists or environmentalism. Within the environmental movement there is a host of ideologies and cross-currents, and there are many classifications of them, which overlap and produce confusion.

The *Oxford English Dictionary* (1972 supplement) has an 'environmentalist' as one who 'believes or promotes the principles of environmentalism' (which is 'a theory of the primary influence of the environment on the development of a person or group') or simply 'one who is concerned with the preservation of the environment (from pollution, etc.)'. Perhaps wisely, the *Dictionary of Human Geography* (Johnston 1981) attempts no more specific a definition - environmentalism consists of 'The ideologies and practices which inform and flow from a concern with the environment'. These broad definitions will suffice here for the present, except that the first part of the OED definition, the use of 'environmentalism' as a synonym for 'environmental determinism' will be dispensed with. Within the definitions as they stand, we can accommodate those who seek environmental reform without corresponding social and economic reform, and those who believe that the former is not attainable without the latter. Although O'Riordan (1981A) believes that the environmentalist movement is broadly reformist, being 'about conviction - conviction that a better mode of existence is possible ... opening up our minds and our organisations to new ideas about fairness, sharing, permanence and humility', there are factions within it which can be justly categorised as being deeply conservative and reactionary, and anxious not to share their portion of resources, dwindling or not. O'Riordan does take up this theme of the contradictory elements within environmentalism. Part of his definition of the movement is that it represents a search for mediation between several contradictory elements e.g. there is individual freedom versus the common good; the protection of national sovereignty versus the need for global awareness; the rights of minorities versus majority rights; the protection of the rights of the present generation versus those of future generations. If we are to solve our environmental problems we have to find ways to resolve the tensions between these opposites, and also between values associated with the desire for affluence and individual material possessions and those connected with social and environmental justice and the non-material, spiritual, sides of our nature.

This last tension is very much to the fore in environmentalist literature. It is startlingly illustrated in the first issue of the *Ecologist*. Volume 1 No. 1 (1970) of this influential 'Journal of the Post-Industrial Age' which headlined its concern with 'Man and the Environment; the Quality of Life; Pollution; Conservation' on its *front* cover, devoted the whole of its *back* cover to an advertisement for Pan-Australian Group of Unit Trusts. This group congratulated the *Ecologist* on its 'timely appearance' and wished it well - adding that 'Pan-Australian makes a habit of leading in capital growth'. More radical environmentalists at least would see here a tension where no mediation is possible.

Not that such tensions are a product only of the 'post-industrial' or even the industrial age. Tuan (1968) has pointed out 'glaring contradictions of professed ideal and actual practice' in terms of environmental attitude and behaviour in the ancient and medieval cultures of China and of the Mediterranean region.

Neither are tensions, ambiguities, or plain and simple hypocrisies characteristics only of 'other people' who express concern for the environment. On an often deeply personal level such concern may, as O'Riordan says, bring us into conflict with ourselves, and draw us into deeper and deeper questions. From a narrow concern about what to do with used bottles, to a deeper concern about the fate of the earth, capitalism and our human souls is but a series of short and logical steps.

1.2 THEMES IN RECENT ENVIRONMENTALIST HISTORY

It is difficult to place the commencement of modern concern about the environment at any specific date, and it is perhaps more useful to speak of landmark publications and events, as will shortly be done. Lowe and Goyder (1983) say that the oldest national environmental group, the British Commons, Open Spaces and Footpaths Preservation Society, dates from 1865. They note the uneven historical development of environmental concern, and identify four major periods when it was being articulated in the West. These were in the 1890s, 1920s, late 1950s and early 1970s - all of them at the end of periods of sustained economic expansion, when people were perhaps more inclined to react against highly materialistic values.

However, it would appear that the last of the four stands apart from the others in at least one sense; that it can be characterised by the occurrence of greater and more widespread concern about a plethora of issues. This concern may have peaked in Britain and the USA in 1967-1974, and dropped off subsequently, but it is by no means at an end in 1984. One of the factors underlying its rise was growing disillusionment among the affluent and educated middle classes (who appear to make up the backbone of the movement - see Cotgrove 1982, Lowe and Goyder 1983) with their very affluence and the materialistic philosophy that supports it. Spiritually, of course, this philosophy is deeply unrewarding, and there is nowadays no compensation for most people in following the traditional Christian religion, though some kind of transcendental experience is clearly sought. This notion of a direct correlation between material well-being and concern for non-material cares and pursuits would help to explain the decrease since the mid-1970s in popular concern for the environment which Sandbach (1980) discerns. For since then economic recession has made many people preoccupied literally with where next year's food, clothing and shelter are to come from, so real is the growth of mass unemployment in W Europe and the USA. Broadly speaking, public concern has now become focussed upon attaining the highest in a hierarchy of national goals depicted by O'Riordan (1981A p19-20). This goal includes economic growth and employment, and 'national security'. At times when these are threatened, they displace concerns which are well down the hierarchy - including environmental quality and ecological harmony, rated only as priority No. 3.

Whether these sorts of generalisations can be applied to whole populations is questionable. For instance, there may be some truth in the idea that environmentalism is at root a protest by the *young* against the values and philosophies of the *older* generation. It is certainly true that many environmental protests since World War Two have involved public demonstrations which have been largely supported by young people. However, as one can see by Allaby's (1971) description of those he termed the *Eco-Activists*, many of the leading figures of the movement have been middle-aged and old. Allaby's subtitle, 'Youth Fights for a Human Environment', may have much truth in it, but it has been the more middle-aged figures like Ehrlich, Commoner, Sagan and Capra, who have led different aspects of the fight. Neither should one forget that Bertrand Russell was well past his three score and ten years when he was so shamefully gaoled for civil disobedience in the anti-bomb protests of the early 1960s.

As suggested, what perhaps distinguished the environmentalist movement of the past two decades from earlier movements was that the former was a mass movement (see Schnaiberg 1980 p368).

Environmental concerns entered mass consciousness through the mass media, frequently at the prompting of mass demonstration of dissent over threats to the environment. To an extent this all-pervasive influence was a result of the mass nature of the threats that were perceived. Pollution was regarded as a large-scale and potentially chronic problem which did not restrict its influence to any specific group of people, but was democratically spread throughout nations and between them.

One can perhaps see a continuum of mass protests over concerns that can be broadly interpreted as environmental. It starts with the big anti-nuclear-bomb protests of the late 1950s in Britain, and brings us up to the current immense peace movement in Western Europe, which is now spreading its way into the USA, Eastern Europe and Australasia. These movements have been about what is clearly the greatest environmental threat, the atomic, hydrogen and neutron bombs - their power to pollute being spectacular and ultimate (Schell 1982). Over the past few years ecology and anti-bomb groups have been steadily coming together in recognition of this fact and in recognition, too, that the arms race symbolises, promotes and is promoted by a whole set of philosophies and socio-economic structures which are inimical to the achievement of of harmony between man and nature, and between men and men. One difference between the earlier and later manifestations of the movement was that in the late 1950s the protestors were calling for the use of 'atoms for peace' (see Pringle and Spigelmen 1982). In the early 1980s relatively few of them make the same call, for it is increasingly recognised that the nuclear energy and nuclear bomb issues are inextricably linked.

In between these movements were the civil rights protests in the early and mid-1960s in the USA, which were, at one level, protests about the quality of the urban environment and way of life of underprivileged groups and were about the blatant maldistribution of resources between these groups and others in American society. During the middle and late 1960s and early 1970s the focus of concern switched to what the USA was doing in Vietnam. There was its direct 'environmental' involvement, in its use of defoliant as a weapon; but also the whole concept of fighting wars in areas remote from the USA came under question. Such wars were ostensibly about 'freedom', but environmentally conscious commentators have seen US and USSR involvement in Vietnam, El Salvador, Afghanistan, Chile, the Near and Middle East and elsewhere as manifestations not of ideological concern but of a worry about safeguarding the *resources* from countries of the Third World which both economies depend on (see Ehrlich and Ehrlich 1972). It is perhaps this concern which preoccupies the two superpowers as they play the game according to the rules of the domino theory.

Contemporaneous with the anti-Vietnam protests in America, and with the campus and street riots in 1968 in Europe and America, was the 'hippy' movement. This was a different kind of protest, being a withdrawal principally by young people from a society and a set of values which they rejected. This movement, too, had its environmental aspects, having very clear philosophical links with the romantic and wilderness movements of the 19th and early-20th centuries (Erisman (1973), London (1969)). Nash (1974, Chapter 13) writes of:

> the emergence in the 1960s of a widespread tendency to question established American values and institutions. The new mood emanated from young people, and in the mid-1960s one-half of the total population of the United States was under twenty-five. These Americans had not been scarred by the Great Depression or World War II. The obsession of earlier generations with success and security seemed sterile and unsatisfying. Rather than accept the world as handed to them and compete for a place within it, many young Americans turned critical and openly rebellious. In sharp contrast to the youth of the 1950s, they felt responsibility for the condition of their world. Impatience with traditional American conventions became commonplace and political activism, especially on behalf of peace and minority rights, became a way of life.
>
> But for others rebellion had a more personal, inward, non-political connotation. 'Freedom', in this sense, meant the opportunity to 'do your own thing'. The unconventional nature of the 'things' that were usually done added up to a revolution in lifestyle. In its vanguard in the 1960s were the 'hippies' (those 'hip' to what was happening), who first became noticeable as a social group in San Francisco's Haight-Ashbury district and New York City's Greenwich Village. Commonly under thirty, poor (by choice), long-haired, bearded, sexually liberated, involved with drugs, folk music, and mysticism, the hippies determined to create a viable alternative to 'square' culture ... It was a shorthand way of indicating that they were disenchanted with prevailing American ideals concerning technology, power, profit, and growth ... Centralization, urbanization, and industrialization appeared as devourers rather than saviours of mankind ... Given this general orientation, the counterculture inevitably discovered wilderness and identified it as something of value. Wild country was clearly on the side - if not the epitome - of nature and naturalness. Certainly wilderness was diametrically opposed to the civilization that the counterculture had come to distrust and resent. Indeed, the American wilderness was a victim of that civilization, a casualty of 'progress', in the same sense as counter cultural values were.

This last reminds us that there was a prominent attempt in most of the 'dropout' communities to re-establish the close and fundamental links with nature and 'mother earth' which were imagined to have existed in pre-industrial rural society. Such links were sought in order to 'recapture' a simplicity and an innocence and gentleness which were perceived to have been lost.

This return to romanticism has also characterised earlier environmentalist periods. Lowe and Goyder (1983) remark upon the attitude change which accompanied the 1890s burst of environmental concern. There was, they say, a late-Victorian reversal of rational progressivist attitudes inherited from the Enlightenment (which held that man could and should improve nature through the exercise of his reason, and via science and technology). Intellectuals like Ruskin, Morris and Mill, who all founded environmental groups, rejected the optimism of economic liberalism and became pessimistic about the prospects for social and economic advancement through laissez-faire capitalism. They were equivocal towards industrialism and, like earlier romantics, saw it as destroying morality and social order, human health and values, and nature. The moral and aesthetic revulsion against the city led to legislation to improve the urban environment, to the garden city movement, and to the escapism of alternative communities which the hippies emulated 70 years later. There was, too, a similar yearning expressed for the 'organic' social bonds which were thought by people like William Morris to have characterised medieval communities. This spirit perhaps lives on in the network of 'alternative' communities which now stretches across Europe and North America.

Although there might in the 1980s have been a fall off in levels of broad popular concern about the environment, nevertheless in some respects environmentalism is still strong and developing. Single issues continue to take much media time and attention, and the more that reactionary Western governments (like those of President Reagan and Mrs Thatcher) intensify policies which are environmentally detrimental, the more will concern grow. Such governments have, for example, used cold-war rhetoric extensively. They have backed up this rhetoric with increased outlay on armaments, but have had also to cope with a burgeoning peace movement whose very growth their rhetoric has stimulated. Large-scale environmental concern has indeed been sustained through the nuclear issue, and the anti-nuclear lobby received a huge stimulus after the near disaster at Three Mile Island in April 1979. Indeed, *The Guardian* was able to report that the nuclear industry in the US was in disarray following the occupation in November 1981 of the proposed site of the Diablo Canyon power plant construction in California. This had led to a deferral of the commencement of work, and the whole of the nuclear industry was in the doldrums, for in the face of similar protests all over the States, and of other, economic, factors, orders for new plant had fallen off considerably.

Other single issues, which have run like festering sores through the late 1960s and the 1970s have continued or re-emerged in the 1980s. In the US the problem of deteriorating air quality and associated acid rain has spread, but gains achieved through legislation have been partly offset by the relaxation of standards which has followed protests from industry - notably the steel and motor industries of the older industrial areas. They have maintained that their international competitiveness has suffered through expensive restrictions on their activities imposed by environmental legislation. Meanwhile, strip mining in wilderness and preservation areas and Indian reservations of the West and South-West has become a big political issue (Bregman 1982). Here again, industry has made some inroads into wilderness legislation, on the grounds that oil shale, uranium and coal must be retrieved as a matter of urgency in the 'national interest'. Edward Abbey's environmental vigilantes (Abbey 1975) in the 'Monkey Wrench Gang' represent in fiction the frustrations and fears of the environmental movement in the face of these threats to fragile desert environments in areas like the Four Corners.

In Britain, somewhat incredibly, there was in 1981-2 an inquiry into whether a third London airport should be built at Stansted in Essex - some 15 years after another public inquiry had concluded that there shold not be such a development. In between, the project has been 'on' and 'off' at different sites, in best soap opera tradition (Buchanan 1981) while other issues, like the extension of the motorway network in Britain generally and around London in particular (see Hall 1981), continue to be subject to the same sporadic attention as a result of vacillations in government policy. They continue, also, to attract notable bodies of protestors. Thus although the memberships of broad environmental movements like Friends of the Earth may have stabilised or faltered (Sandbach 1978A), concern over specific issues, and the constant re-emergence of those issues, keeps environmentalism of one sort or another alive and kicking.

1.3 MAJOR CONCERNS OF MODERN ENVIRONMENTALISM

During the heyday of modern environmentalism, much of the attention fell on to the old issue of the population/resources ratio - the issue which had preoccupied Thomas Malthus over a century and a half earlier. Like Malthus, 'neo-Malthusians' attempted to take a global perspective, speaking in broad terms about world-wide exponential population growth, and potential incapacity to feed millions of new mouths by the turn of the twentieth century. In a manner which was partly reminiscent of Malthus, Garrett Hardin soon brought the concern round to one of ethics and morality. He did not see that the 'population problem' could be solved without repudiating certain fundamental ethical beliefs. He drew attention to the (inescapable) biological law of *carrying capacity* in his 'Tragedy of the Commons' (1968) allegory. In this story a piece of common land was grazed by a certain number of cows. It was fully stocked, being in an equilibrium state whereby maximum productivity was achieved without a deteioration in the quality of the sward. Hardin pointed out what happened when one more cow was added by a herder who saw how well all the other herders were doing and tried to emulate them. The addition led to overgrazing and destruction of the whole commons. All had suffered from the action of one person - an action which had come about through ignorance rather than premeditated selfishness. But Hardin did not draw from this the moral that mass education for environmental responsibility would avoid such ecological problems. He felt that only one person needed to act irresponsibly to bring disbenefits to all, and that that one person would always exist - so that common ownership of resources based on a community of interest would never work. Only enlightened private ownership and restriction of access for the mass of people, or restriction of the size of the masses, could protect the commons.

This has developed into a parallel central question for environmentalists of both left and right political persuasions. It is the question of the legitimacy of the pursuit of private interests for *individual profit* from resource exploitation, where the *losses (environmental and social)* are borne by all. One can see this issue in the motorways controversy in Britain, where calculations of the disbenefits of new roads seldom refer very thoroughly to the environmental and social losses, which are regarded even less as losses to be borne by the British Roads Federation, the private organisation which will be the chief financial beneficiary.

Those on the left might argue that the question should be resolved by redistributing the profits from resource exploitation, or the resources whereby to profit; Hardin argued against such a redistribution - on grounds not of selfishness, but of pragmatism in deference to biological laws. Thus, in another allegory (1974) he posited 10 men in a lifeboat adrift in a sea full of drowning men calling out for help. The lifeboat has supplies for only 10 men. To share them between 11 or more by pulling someone into the lifeboat would mean that no-one got enough, and all would starve. Thus it would become impossible to exercise compassionate sharing because of the universal application of the carrying capacity law. This message soon became transposed to the Western World/Third World debate, and it became an argument against the provision of food aid to the Third World.

Those Westerners who were in the lifeboat, with its adequate food supply, *could not* for practical reasons help those (of the Third World) who were drowning, and it would be irresponsible to do so. For Hardin believed that 'Any nation that asserts the right to produce more babies must also assume the responsibility for taking care of them'. This was the nub of the problem, not the inability to produce food. 'The problem is too many people. The food shortage is simply evidence of the problem' (The Environmental Fund 1977).

This 'simple' evidence, with its portents of doom, lay behind biologist Paul Ehrlich's public appeal and impact in the period. In 1969 he spoke to the Institute of Biology in Britain and forecast that because population growth would outstrip resource availability there would be a radical lowering of the *quality of life* (a key concept in environmentalism) in the West. Ehrlich was also an alarmist, maintaining that because of this and because of the build-up of toxic pollutants such as organochlorides and metals in food chains 'We' (i.e Ehrlich's generation) 'are already dead'. He saw population growth as the root cause of environmental deterioration, and founded the Zero Population Growth Movement. This contended that it was already too late voluntarily to stop breeding above replacement rates, and that coercion and even anti-population-growth laws were needed to curb the excesses of some groups and nations. Barry Commoner, in 1971, called this the 'New barbarism of the lifeboat ethic'.

Commoner thought that it was not population growth but the *nature of technology* which was to blame for the world's environmental predicament. Modern technology concentrated increasingly on substituting synthetic and non-*biodegradable* products for 'natural' products. This led to a surfeit of toxic residuals in the biosphere, lithosphere and atmosphere. Commoner later became a candidate in the 1980 US Presidential election and his platform had developed these early ideas on the nature of technology into questions of who *controlled* it. He stood on an anti-multinational-corporation ticket, on the grounds that the multinationals were too powerful and behaved in an unsympathetic way to the needs of man and nature. (Predictably, he failed massively to get elected.)

In developing his ideas this way, Commoner displayed a tendency which several environmentalists showed during the 1970s. Their ideas developed in a politically leftward direction, and they focussed on *fundamental social relationships and processes* which were seen to cause the phenomena with which environmental deterioration is associated rather than simplistic explanations like pure population growth. Thus, Michael Allaby (1970) put the blame on population growth at the beginning of the decade. But ten years later he considered that this had been wrong. *Maldistribution of resources* constituted the real root cause of our environmental plight, he maintained (Allaby 1980). This position has not, however, been consistently taken on the left. In 1982, for example, novelist John Fowles, in an article proclaiming his socialism, also said 'All international and most civil conflict is at root caused by this terrible bane (overpopulation) with its totally unneeded pandemic of hands to employ, wages to find, mouths to feed' (Fowles 1982).

Others have been more sanguine about modern technology, and the science which has produced it. At one highly optimistic extreme, Kahn (1980) predicted that with the proper application of 'free' market policies, unbridled economic growth will be possible and that technology will be able to overcome environmental difficulties associated with such growth. This *technological optimism*, which Cotgrove (1982) calls 'Cornucopian', has been the general tone of big business interests throughout the past 15 years when confronted with their own environmental 'mess', and it has also been a theme pursued by those with less obvious vested interests in environmental exploitation. Thus the enormously influential Reith Lectures on 'Wilderness and Plenty' given in 1969 by the naturalist Frank Fraser-Darling (1971) found plenty to be pessimistic about, but they nonetheless, after some heartsearching, ended on an optimistic note which did not totally reject the marks of material progress, even though the values on which it was founded might have to be modified.

In Britain these Reith Lectures led to a welter of publicity for environmental issues, such as that of London's third airport and the government decisions to have it and to site it at Stansted despite a weight of influential counter-opinion. The general question soon arose of just how *democratic* was and is the decision-making process over land use/environmental matters. Because of the inept government record over the third airport, an 'independent' inquiry was set up in 1969 to consider the 'timing of the need' (not the need itself) for an airport and where it should be sited. Lord Chief Justice Roskill, the head of the inquiry team, made much of its intended objectivity - and to achieve it the technique of economic *cost-benefit analysis* was extensively employed. The attempt was made to convert all the factors in the equations for and against specific sites into money terms. Much has since been written on the absurdities of such an approach. The value which the Commission put on the fine Norman Church at the village of Stewkley, which would have to be dismantled if the inland site at Wing in Buckinghamshire was to be chosen, was a ludicrous £11,000 (being the insurance value) - and this church has since become the symbol of the virtue and potency of allowing *intangibles* to remain intangible, as against attempts at 'hard-headed' economic appraisal of some of our environmental assets. However, the latter approach has generally persisted in government ministries and agencies. This, even though *economic values* were ostensibly rejected for *environmental values* when the eventual Roskill findings in favour of Wing were put aside by the government for the less 'economic' coastal site at Maplin Sands - where potential environmental damage was reckoned to be far less extensive.

The rejection of an inland for a more expensive coastal site was proclaimed by some as a great victory for environmentalism, in which the values attached to (economic) standard of living were rated as secondary to those of (environmental) quality of life. Others, however, believed that the real environmentalist victory came in 1974, when the project was dropped altogether (Pepper 1980). They were thus dismayed when it was revived in 1979.

The importance of environmental values was apparently enshrined in the public institutions of the US and Britain, when the Council for Environmental Quality, Environmental Protection Agency and Department of the Environment (soon to be known unaffectionately among environmentalists as the 'Department of Environmental Disaster') were founded in rapid succession in 1969-70. Such quick victories for the environmentalists were rather hollow for, as has been noted above, economic growth aims soon became once more paramount after the Arab Oil Crisis hit the West in 1973, and during the subsequent economic recession. It seemed that almost overnight environmental gains were set aside when, in 1974, the building of the Alaska pipeline was sanctioned after environmental groups had apparently stopped it. Similarly, planning procedures were 'streamlined' in Britain in order to speed the building in beautiful Scottish lochs (like Kishorn) of oil platforms for use in the North Sea.

1.4 THE 'LIMITS', THE 'BLUEPRINT' AND 'SMALL IS BEAUTIFUL'

The principal ideas of modern environmentalism were expounded in three landmark publications of the environmental heyday period. These were *Limits to Growth*, *Blueprint for Survival* and *Small is Beautiful*. The first attempted to enunciate the range and nature of the fundamental problems, the second tried to focus on the kind of changes which would be needed to solve them, and the third attempted to mingle this practical approach with a consideration of philosophical roots of the problems. Although all three works have made an abiding mark, the last has perhaps been most deeply influential.

The *Limits to Growth* team at the Massachusetts Institute of Technology attempted to model nothing less than the world on their computer. Their complex model was used to predict in a detached, rational and 'scientific' fashion what would be the effect of continuing the 'physical, economic and social relationships' which have 'historically governed the development of the world system'.

The model was used to make graphs of parameters such as 20th-century resource use and depletion, population growth, pollution generation, income and food per capita, etc., from a global perspective. They were projected from 1900 to 2100AD. Assuming the continuation of present 'relationships', this exercise found that exponential depletion would take the graph of resource availability below the exponentially-rising graph of population levels in the near future (i.e. around the turn of the 20th century) to produce eventual population collapse and crisis. This 'standard' run was modified by feeding various assumptions of change into the world model, one by one, and then all together (the assumptions included the full use of resources and a recycling of 75% of them; the doubling of agricultural output; the reduction of pollution levels to 25% of those of 1970; and making available effective birth control methods to the whole world). In all cases, however - through world population growth, resource depletion or pollution accumulation - industrial and/or agricultural output declined, followed by a decline in population. The earth's carrying capacity was exceeded, with disastrous results, though with all the assumptions operating the date of this *'overshoot and collapse'* could be postponed for nearly a century.

Only a 'stabilised world model', however, avoided overshoot and collapse altogether, with an emphasis on simultaneous population and pollution control, recycling policies and an abandonment of economic growth as an overall world policy (though the study did say that the present maldistributin of resources between rich and poor nations must be eliminated). The message was not merely of the need for zero population growth, it also advocated *zero economic growth*, if world 'eco-catastrophe' were to be averted. The study made a huge impact on the world's media when it was published in 1972 - though, as mentioned in the Preface, a report which was similar in tone and findings, the *Global 2000 Report to the President*, caused hardly a ripple when it was issued in 1980. *Limits to Growth* was criticised on many grounds. Very many of the critics suggested that imperfect or wrong data, based on incorrect assumptions, had been fed into a very unrealistic world model. However, the reply to such criticisms had a certain validity - precisely *when* the overshoot and collapse would occur was of secondary importance. What really mattered was the general conclusion that a global economic and population 'steady state' had to be achieved if the collapse were to be avoided altogether. However, others voiced a deeper concern. The capacity of human beings to respond with a measure of inventiveness, improvisation and independent thought to the constraints placed upon them by the environment had not, they said, been adequately taken into account. Man was not a mere 'biocybernetic' device - he was not a machine capable only of responses which were quantifiable, predictable, and strictly limited in scope. O'Riordan put it thus: 'To programme values and political responses into a computer somehow denigrates the spontaneous nature of the human mind and its democratic ideals' (O'Riordan 1981A). Maddox, a technological optimist, felt that the report was counter-productive. With its gloom-and-doom overtones, it might 'sap the will of advanced communities' to deal with the problems it identified (Maddox 1972).

Golub and Townsend took a more cynical view of the whole study - or at least of the way in which selected aspects of it had been emphasised by the 'Club of Rome', the group of leading western industrialists who had sponsored the work in the first place. They had, it was maintained, hoped to create a climate of despondency and foreboding in which people would be willing to sacrifice national aspirations for the 'greater global good'. This would mean greater acceptance of the notion of international bodies exerting control over national economies - and within such a supernational framework the multinational corporations which were represented by the Club of Rome would be freer to act in their own interests, unfettered by nationally-imposed constraints on their freedom.

The Ecologist's Blueprint for Survival also spent some time discussing the nature of man's environmental predicament. But essentially it aimed to dispense with criticisms to the effect that environmentalists were generally too idealistic and impractical, failing to say *how* they would achieve their goals. *Blueprint* set out a somewhat detailed programme for future British society to model itself on - accompanied by a timetable for its achievement. In many ways the publication also set out a future geography of Britain, with its emphasis on changed landscapes and a very different urban and regional organisation, based on the twin ideals of decentralisation and smallness of scale. It laid down fundamental goals in which human activity should involve minimum ecological disruption and the maximum conservation of energy and materials. It should also be based on a stable population at about half its present size, since this was commensurate with notions of national self-sufficiency. Since achieving these goals would involve enormous socio-political problems - including the fact that people would generally be materially poorer - the *Blueprint* aimed to create through education and other means a social system in which the goals would be willingly accepted by the majority of people. Their chief compensation for a deterioration in material standards would be an enhanced quality of life - the concept involving greater harmony between man and nature, and between people. Urban and rural environments would be improved; the nature and quality of work would change; with the rejection of impersonal production based on large-scale production lines and division of labour, the creative element would be reintroduced. Through education, value systems would be reoriented to place spiritual and emotional aspects of life in high esteem. People would have deeper relationships with their fellows. They would spend time in creative work and leisure, developing their talents to the full. Their sexuality would no longer be suppressed, and neither would racial minorities or women be subordinate elements in society. Other features would include production through 'appropriate' technology, and extensive recycling of raw materials. Deleterious technology would be phased out in preference for less deleterious technology (e.g. organic fertilisers would replace inorganic fertilisers; natural pest controls would be substituted for artificial ones) and there would be vigorous alternative 'soft' energy and energy conservation programmes.

All of this would be more feasible in *small communities*. In them, also, people would be more fulfilled through being directly in control of their own society - that is, they would have a clear influence on government, which would be largely local. Small communities would also, it was thought, have a minimal adverse impact on the 'natural' ecosystem. The *Blueprint* was specific - it spelt out by what dates different stages in its programme would be achieved, and it detailed the regional structure of Britain, with the size of its regions. It placed emphasis on personal happiness, fulfilment, control over technology and society, education, the importance of creativity and art, and on spirituality. These emphases seemed to be advocated primarily for *ecological* rather than *humanitarian* concerns. Society must be organised in this way because of the operation of ecological laws and not necessarily because of a major concern for social justice. One feels that although the end result might be a socially more just society than the present one, such a result could be almost incidental to the main aims, and it did not follow that there would be such a result. Thus when the *Blueprint* was translated into the fiction of *Ecotopia* (Callenbach 1978), it was a moot point as to whether this mythical country, run in 1999 on ecological principles (and formed by the secession from the USA of Oregon and N. California), was a socialist utopia or a fascist dystopia (see Chapter 7).

When one attempts to map out a geography and sociology of the *Blueprint*-style ecotopia, it appears to differ little from that which is also to be derived from the work of Fritz Schumacher. The emphasis of *Small is Beautiful* however, was more on abstract ideas and, as already indicated, one of the author's main tasks was to expose the nature and deficiencies of the currently-held philosophies which govern our relationship with nature. He examined the values which stem from such philosophies, illustrating how they translate into economics. His emphasis tended to be on the idealistic notion that values shape economics (see Chapter 6), and he drove us towards a solution to the environmental 'crisis' which hinged upon the need for a changed value system in the West.

Schumacher encouraged us to see the possibility of alternative forms of economic systems to our own, and that the 'laws' of economics which we are used to thinking of as universally-applicable are not in fact so. Given different national values and goals, different forms of economics would obtain, such as that which he termed 'Buddhist economics', and therefore different sets of economic 'laws' would apply. This is important, because if we wish, for example, to determine whether it is 'economic' to employ people or machines to do certain types of job our answer will be different according to what economic and value system we subscribe to. If we see the *goal* of our system as that of capital formation through the pursuit of profits, then it is usually 'economic' to replace labour with machinery in doing work. But if the goal of our system is to produce happy and fulfilled people, then it makes *economic* sense to support an organisation of work which creates jobs but does not necessarily maximise profits. Schumacher was much concerned with work, and the need for it to be fulfilling and creative - i.e. one of the key components of the *Blueprint's* 'quality of life'. To improve the quality of work as part of an improved quality of life he proposed to reject the notion that 'high' (i.e. sophisticated and capital-intensive) technology is of merit for its own sake. He sought to encourage the development of simple machines which could be accessible to - and owned by - the majority of people, and which could be mixed in with manual labour to derive a partially-mechanised production process that would generate work. Thus the division-of-labour/production-line philosophy of classical economics would be deliberately destroyed. Schumacher's ideas have been put extensively into practice in the Third World as well as in Europe (see McRobie 1982), and he elaborated upon them in *Good Work* (1980), published after his death.

1.5 ANALYTICAL FRAMEWORKS

Many classifications of environmentalists and environmentalism have been attempted. We shall refer here to three different sorts of classification.

The first classifies ideological themes or 'lines of thought' 'which arose at the birth of the conservation movement'. These constitute the belief systems and arguing positions of environmentalists, and are broadly resolvable into *'ecocentric'* and *'technocentric'* modes (O'Riordan 1981A). The second is a classification of social science theory and explanation - explanation, that is, of how society works, and therefore about how decisions are made on environmental issues. It distinguishes between, on the one hand, *functionalist/pluralist* perspectives, and, on the other, *Marxist* perspectives. The classification also provides a vehicle for describing and explaining the growth of the modern environmentalist movement. It has been developed for environmentalism by Sandbach (1980). The third is a classification of underlying philosophical positions on the man-nature relationship. Using familiar philosophies, it applies them in a manner suggested by D.E. Cosgrove and used by him in his teaching at Oxford Polytechnic. The fundamental distinction here is between *determinist* and *free will* positions.

It is important to note several things about the use of these classifications. First, they are of particular groups of people and their perspectives on the environment at different levels of abstraction. Thus an ecocentric may display a number of deterministic attitudes in his statements, which could also indicate a pluralist view of society. Second, they are not hard and fast or mutually exclusive classifications. Thus, some groups show characteristics of both ecocentrism and technocentrism, while a technocentric viewpoint will probably have elements of both determinism and free will philosophies in it. Also, many individuals will show characteristics of both - thus, 'we should avoid the temptation to divide the world neatly into an ecocentric camp of environmentalists and a technocentric camp of manipulative professionals and administrators. In real life the boundaries are much more blurred. There is every reason to believe that each one of us favours certain elements of both modes, depending upon the institutional setting, the issue at hand, and our changing socioeconomic status' (O'Riordan 1981A p2).

Third, in the case of the philosophical classification, we are dealing with a continuum, of which the two 'categories' of determinism and free-will represent extreme opposites. We expect, therefore, those extremes to find expression infrequently - rather it is a matter of identifying where people and groups lie within that continuum, and anticipating that they will commonly incline towards one or the other end without 'reaching' it.

Of the three classifications, we shall describe the first in a little detail here and now, since it is easy to see its applicability to the kind of issues and publications which have been discussed above. Also, armed with this conceptual framework, the reader should be able to read or listen to a debate on any modern environmental issue and readily recognise whether he is receiving an ecocentric or technocentric view. The second and third classifications may be more readily understood after some discussion of the historical antecedents of some of the lines of thought in environmentalism, and we therefore intend now to describe them perfunctorily, but to defer any lengthy discussion of them until later. The eventual treatment of them may also be partial, concentrating on those aspects which do not find frequent and extensive expression in everyday debates in the media, and/or which are less easy to understand and recognise. Thus, while many environmentalists and perhaps most technocentrics among them would agree with the conventional view of Western society as a democratic structure, analysable within a pluralist framework, the Marxist interpretation of the same society is given less common currency and will be less familiar - especially in a readily-understandable form. We hope to remedy this situation a little, and if it necessitates affecting a bias in a reverse direction we will do so.

1.5.1 Ecocentrism and Techocentrism

Ecological and technological environmentalism or ecocentrism and technocentrism, are most clearly understood in terms of O'Riordan's (1981A) original description. He saw the *ecocentric* root of modern environmentalism as 'nourished by the philosophies of the romantic transcendentalists of mid-nineteenth-century America'. These advocated a democracy among God's creatures, such that nature was respected for its own sake, above and beyond its usefulness or relationship to man. Therefore man had a moral obligation towards nature 'not simply for the pleasure of man, but as a biotic right'. This essentially non-rational and even emotional belief, to which the epithet *bioethic* is given, means that if man were to be wiped off the face of the earth tomorrow there would still be a purpose and meaning in the continuance of life on the planet.

However, while man might not be necessary to nature, the reverse is not true, for, says the ecocentric, 'natural architecture has a grandeur which both humbles and ennobles man and stimulates him to emulate it. Wild nature ... is an integral companion to man' (O'Riordan p4) necessary for his emotional, spiritual and physical wellbeing in the face of pressures from sophisticated and artificial urban living. While there is not necessarily any biological or economic justification in the bioethical value system ecocentrism is wide enough to embrace also the views of those who argue for nature on more pragmatic and rational grounds. This argument, from an essentially scientific *ecosystems* perspective, puts man *within* nature, as part of natural ecosystems. Consequently, anything which man does affects the rest of the global system and reverberates through it - eventually back on to him. So, for his own sake, he should not plunder, exploit and destroy natural ecosystems - because in so doing he is destroying the biological foundation of his own life. Man is seen as subject to biological laws just as much as is the rest of nature, and so he must contribute to the stability and mutual harmony of the ecosystems of which he is a part. The biological law of carrying capacity has already been mentioned in this respect, but other 'laws' governing population size and dynamics, or laws of thermodynamics or laws governing systems behaviour (e.g. diversity equals stability) are held to apply also to social and economic man. Indeed, the whole paraphernalia of systems terminology is applied by the ecological school - sometimes to extremes which are faintly ludicrous (e.g. Goldsmith 1978).

Limits, self-reliance, self-sufficiency, small-scale production, low-impact technology, recycling, zero population and economic growth - these are all key words in the standard ecocentric vocabulary, which is liberally sprinkled through the three landmark publications described above. The *Blueprint* and *Small is Beautiful* are undoubtedly 'ecocentric' in outlook, though *Limits* has technocentric as well as ecocentric characteristics.

All the above may be qualified by reference to O'Riordan's (1977) political sophistication of his classification. Thus, in discussing the ideological cross-currents of environmentalism he distinguishes between 'conservative ecocentrist' and 'liberal ecocentrist' ideologies. While the former embraces the morality of limits and of lifeboat ethics, and its adherents belong to the no-growth school and to the ecological planners and amenity protectionists, the adherent of the latter is classifiable as a 'radical ecological activist' - i.e. an 'environmental educator' or citizen, who generally 'seeks fundamental changes in the values, attitudes and behaviour of individuals and social institutions through example and enlightenment, not by revolution or chaos'. Cotgrove (1982) makes a similar distinction, between 'new environmentalists' and 'conservationists' - the latter revealing themselves, according to his survey data, to be politically more to the right than the former.

While he will frequently recognise the existence of environmental problems and desire to solve them, the *technocentric* differs from the ecocentric in how he would approach such problems, and in his basic ideologies. He is identified by an apparent undiluted rational, scientific approach, which particularly translates itself into an economic rationality founded on the neo-classical school, There is, too, a belief in the ability and efficiency of management in solving problems by the use of 'objective analysis' and recourse to the laws of physical science - the natural authority of which is extended to economic 'laws'. This management includes management of the environment - and of men, for unlike the ecocentric the technocentric disavows public participation in environmental and other decision-making in favour of accepting as authoritative the advice of (scientific and economic) 'experts'. Although this is ostensibly a rational mode, such rationalism, says O'Riordan, may be stripped away to expose a raw and sometimes irrational faith - a faith in the idea of progress as expressed in, and equivalent to, material advancement, in the superiority of 'high' over 'lower' technology, in the sustainability of economic growth, and in the ability of advanced capitalism to maintain itself. Frequently those who express such faiths have much to gain materially by their application. And their resultant undetached and unobjective position manifests itself in an irrationality which clearly transgresses the technocentric's own terms. Thus a truly 'objective' and 'expert' cost-benefit analysis would probably have grounded the Concorde project before it ever left the drawing board. It would probably have stopped the nuclear power plant building programme of the British Conservative Government which came to office in 1979, for many economic forcasts of demand for fast travel and for energy showed that both programmes would be redundant in the face of Britain's declining future needs.

However, the technocentric is a complicated animal, for if irrationality lies behind the rational facade, so too, according to O'Riordan, does a lack of confidence lie beneath the authoritative expert aura which he exudes. For if one 'strips off the veil of optimism' one can reveal underneath an inherent and disquieting uncertainty, prevarication, and tendency to error. Thus, the management of British Nuclear Fuels Ltd. argued vehemently at the 1978 Windscale Inquiry into the reprocessing of atomic waste that adequate and stringent safety precautions were taken at the Cumbrian atomic plant where the reprocessing was intended, against the leak of atomic waste. Yet, two years later, a report was finding evidence of managerial incompetence over radioactive waste which had leaked some years earlier into the soil surrounding the plant, while in 1983 Government legal action was contemplated against the management because of new leaks. And the story of the accident at the Three Mile Island nuclear power plant in Pennsylvania in 1979 is studded with examples of the technocrats' prevarication and error to a remarkable degree (Stephens 1980).

The technocentric mode does not necessarily declare itself in favour of environmental degradation: usually the reverse. But it holds that this is a matter for efficient *environmental management*; for 'cleaning up' after the mess made by necessary modern industrial processes, the abandonment or amelioration of which would be regarded as a reversion to some form of primitive barbarism. And in any conflict between the demands of economic man and environment, where the interests of the two were not reconcilable through management, economic man would win the day: there is no bioethical sense of nature's rights. Unlike ecocentrics, technocentrics usually address themselves to short-term planning, rarely taking a perspective on the decades and centuries beyond 2000AD.

Again, there is a conservative/liberal division among the technocentrics. The former espouse the growth morality, and are technological, managerial and political optimists *par excellence* (Cornucopian); the latter are cautious reformers and social democracts of the 'materials balance' economic school, i.e. the 'accommodators' (O'Riordan 1981B).

The basic ideas in this classification have subsequently been modified by O'Riordan, and shorn of their political 'conservative/liberal' nomenclature, but they remain recognisable (Figure 2). Indeed, as already suggested, the two modes are eminently recognisable in almost any public debate about modern environmental issues. Thus, Del Sesto (1980) analysed the testimonies given by members of the pro- and anti-nuclear lobbies to the 1973-74 Congressional hearings on the status of nuclear reactor safety held by the Joint Committee on Atomic Energy. He abstracted from the statements various characteristics which he summarised as the ideologies of the two sides. The pro-nuclear lobby (Figure 3A) is instantly recognisable as technocentric, while the anti-nuclear (Figure 3B) is clearly ecocentric.

1.5.2 Functionalist/Pluralist and Marxist Perspectives

These perspectives on society are distinguished by Sandbach (1980) according to whether they place fundamental emphasis on the achievement of social change through continuous reform of the present socio-economic system (the former) or by overturning the system altogether (the latter). Functionalism is described as a social science equivalent of the systems approach in natural science, whereby social problems are seen to develop as a result of systems imbalance or malfunction. As a result of this, stresses occur and social change takes place in response to them. The stresses arise from the (political) activities and pressures exerted by a plurality of pressure groups, many pursuing their own vested interests. Because society is essentially democratic, each group has opportunities to articulate its position and the resultant socio-economic change will be such as to take account of, and perhaps mediate between, different positions. So the outcome is a society which evolves in an essentially gradual way through democratic competition between different groups. Decision-making about the environment and environmental reform reflects, therefore, the plurality of interests surrounding such issues, and thus some sort of compromise between ecocentric and technocentric modes is usually likely. As Sandbach says (p29) 'In the functionalist and ordered world there is

= the will of all?

ENVIRONMENTALISM

ECOCENTRISM ← → TECHNOCENTRISM

Deep ecologists	Self-reliance Soft technologists	Environmental managers	Cornucopians
Intrinsic importance of nature for the humanity of man	(1) Emphasis on smallness of scale and hence community identity in settlement, work and leisure	(1) Belief that economic growth and resource exploitation can continue assuming:	(1) Belief that man can always find a way out of any difficulties either political, scientific or technological
Ecological (and other natural) laws dictate human morality		(a) suitable economic adjustments to taxes, fees, etc.	
Biorights—the right of endangered species or unique landscapes to remain unmolested	(2) Integration of concepts of work and leisure through a process of personal and communal improvement	(b) improvements in the legal rights to a minimum level of environmental quality (c) compensation arrangements satisfactory to those who experience adverse environmental and/or social effects	(2) Acceptance that pro-growth goals define the rationality of project appraisal and policy formulation (3) Optimism about the ability of man to improve the lot of the world's people
	(3) Importance of participation in community affairs, and of guarantees of the rights of minority interests. Participation seen both as a continuing education and political function	(2) Acceptance of new project appraisal techniques and decision review arrangements to allow for wider discussion or genuine search for consensus among representative groups of interested parties	(4) Faith that scientific and technological expertise provides the basic foundation for advice on matters pertaining to economic growth, public health and safety
(4) Lack of faith in modern large-scale technology and its associated demands on elitist expertise, central state authority and inherently anti-democratic institutions			(5) Suspicion of attempts to widen basis for participation and lengthy discussion in project appraisal and policy review
(5) Implication that materialism for its own sake is wrong and that economic growth can be geared to providing for the basic needs of those below subsistence levels			(6) Belief that all impediments can be overcome given a will, ingenuity and sufficient resources arising out of growth

FIGURE 2   Ecocentrism and Technocentrism (source: O'Riordan 1981B)

little real conflict of social interest, but only deviation from the consensus. Social movements within the system arise as a consequence of tensions and deviations from the normal ordered state of affairs.' The rise of environmentalism itself can be interpreted within this framework, and one can envisage a future society in which some at least of the environmentalist worries and concerns have been accommodated and a new relative consensus, or dynamic equilibrium, has been reached.

When the functionalist model is adapted to a pluralist framework, one sees that environmental issues may arise as a result of conflicts of interests between different groups. But because there are many such conflicting interests and pressures:

> there is no bias towards any ruling group. Any attempt at domination would be checked by the majority through various democratic mechanisms such as elections and many other constitutional and legal safeguards. Pluralism therefore accepts that there must be a consensus of public opinion for social change to take place ... this reliance upon a common accepted interpretation ... makes pluralism particularly compatible with functionalism (Sandbach 1980 p31).

Sandbach goes on to suggest that just as neo-classical economics tend to be part of the conventional wisdom, as reflected in and reinforced by the majority of the media, so too do the assumptions of pluralism underlie the 'dominant paradigm', as Cotgrove (1982 p27) calls it, in society. The dominant paradigm is dominant not necessarily 'in the statistical sense of being held by most people, but in the sense that it is the paradigm held by dominant groups in industrial societies, and ... that it serves to legitimate and justify the institutions and practices of a market economy'.

The less-commonly-heard Marxist perspective on social and environmental change will be described more fully in section 6.5, but unlike pluralism it emphasises the importance of *conflict* between vested *class* interests. For example, environmental concerns could be said to reflect the clash of interests which is growing between those who have ownership of resources (for example, raw materials, or rural amenity) and those who increasingly want them or access to them for themselves - that is, between those 'in the lifeboat' and those drowning in 'the sea' of scarcity. The former could be interpreted as the ruling and capital-owning classes and the latter as those who have only their labour to sell. The model could be applied in the context of the economically 'developed' countries, or internationally. In the latter circumstance those defending their environmental and resource interests become the developed countries while the countries of the Third World represent those increasingly threatening these interests by their demands for resources and higher material standards. Since such conflicts inevitably arise out of the inherent contradictions within the capitalist economic system, they cannot be resolved through pluralist-style modificatory reform of the system, but only by a total overthrow of that system.

Furthermore, whereas pluralists have faith in democracy, and the

FIGURE 3A. MAJOR ELEMENTS OF THE PRO- NUCLEAR IDEOLOGY

The Promise of Peaceful Applications
   The use of nuclear power raises the standard of living
   The use of nuclear power increases economic growth
   The use of nuclear power promotes fiscal well-being
   The use of nuclear power provides unlimited energy
   The use of nuclear power will solve the energy crisis
   The use of peaceful nuclear power will deemphasize military uses

A Faith in Science and Technology
   Political and value issues associated with nuclear power are amenable to scientific and technical solutions
   Science and technology can solve all practical problems associated with nuclear power

Impeaching the Opposition
   The opposition misrepresents or distorts the facts
   The opposition uses exaggeration and scare talk
   The opposition employs personal, *ad hominem* attacks
   The opposition has no credibility

FIGURE 3B. MAJOR ELEMENTS OF THE ANTI-NUCLEAR IDEOLOGY

The Legacy to Future Generations
   The use of nuclear power is morally indefensible
   The use of nuclear power will lead to a totalitarian police state
   The use of nuclear power will promote sabotage and diversion of fissionable material, resulting in restrictions on individual freedoms
   The use of nuclear power will result in proliferation and eventual war
   Science and technology cannot solve all the problems associated with nuclear power

Anti-centralization and Political Accountability
   Nuclear power decision-making should be decentralized
   Nuclear power decision-making lacks due process
   Nuclear regulators should be made more politically responsible
   Regulatory proceedings should be more responsive to citizen groups and public participation expanded
   Decentralized, low-technology alternatives are preferable to nuclear power

Impeaching the Opposition
   The opposition distorts and suppresses the facts
   The opposition employs personal, *ad hominem* attacks
   The opposition is involved in collusion and conspiracy against the public
   The opposition has no credibility

FIGURE 3   Nuclear Ideologies (source: Del Sesto 1980)
             (Copyright John Wiley and Sons Inc.)

power of argument to change society, Marxists see that the ruling classes (capital owners) have the power and means to distort the argument by constraining the influence of the minority who argue against the dominant paradigm, and by curtailing the opportunities of the majority to learn that there *is* a feasible alternative to the dominant paradigm. This constraining can be done in state capitalist countries (like the USSR) by direct and obvious repression and persecution (viz. the confinement in psychiatric wards of members of Moscow's independent peace group reported in the *Guardian* 1 September 1982). In Western economies, dominated by the interests of private capital, more indirect means are relied upon, such as capitalist ownership and control of the press, radio and TV, or influence through government on the educational curriculum (viz. the Department of Education and Science's strictures against the Open University in August 1982 for including 'Marxist bias' in certain social science courses - see *Times Higher Education Supplement* No 313). Thus, in the Marxist view it is pointless to have faith in the power of argument - or to believe (as ecocentrics commonly appear to do, see Preface, section 1) that if only people knew 'the facts' about man's abuse of nature, then such abuse would stop. For the only people who will listen to the argument are those whose class interests the argument will serve. Thus, coal miners' representatives are against nuclear power, and atomic scientists are pro-nuclear. Both claim to be 'objective' in their approach, and assert that their arguments are superior to those of the other side. The Marxist onlooker (and many others) would claim, however, that in their own postures both sides are obviously attempting to defend their own vested economic interests.

1.5.3 Determinism and Free Will

One can analyse environmentalism in terms of the sociological composition and ideologies of the groups involved, and of the way in which society responds to growing awareness of environmental difficulties and of what processes cause these difficulties - as O'Riordan, Sandbach, Schnaiberg, Cotgrove and others do. However, underlying all these forms and levels of analysis and the ideologies they reveal are also fundamental philosophical positions, as has been suggested at the beginning of this work. These positions will be considered using largely the *determinist-free will* framework. An understanding of this will hinge on the material which follows, but a brief outline here of what is involved at these two extreme ends of a philosophical spectrum may be helpful.

. The tradition of environmental determinism, which has been so strong in, for example, geography and ecology, seeks to establish scientifically the nature of the relationships between man and nature, stressing the view that this relationship is one in which environment 'controls the course of human action' (G.R. Leuthwaite, quoted in Johnston 1981). The environment - usually defined in terms of physical factors like (especially) climate, soils, topography - is believed to *influence to the extent of substantially controlling*, aspects of human behaviour (individual and group), economic activity and social organisation, and even physiological characteristics. Glacken (1967) traces the development of this philosophy from its enunciation by Hippocrates in the 5th century BC, in his discourse on *Airs, Waters and Places*, to its pre-eminence in 19th-century thought. In the former, the influence of topography and climate on human health was stressed. By the 19th century the philosophy had diversified to consider the limits on human population size which were imposed by limited availability of natural resources (particularly for producing food) and more generally, through the influence of Darwinists, how natural laws operate to differentiate between men. The modern ecological version of the philosophy clearly places man within nature, and as a part of it, being dependent on it and subordinate to its laws.

By contrast, a free will philosophy such as *existentialism* is 'a body of philosophical doctrine that dramatically emphasises the contrast between human existence and the kind of existence possessed by natural objects. Men, endowed with will and consciousness, find themselves in an alien world of objects which have neither' (Bullock and Stallybrass 1977). Existentialism emphasises human personality and free will. Humans are distinguished from natural objects by their individual and unique existence, so they cannot be explained in terms of any scientific system of universal laws, being essentially *free to choose* what they will do. Such freedom clearly involves the freedom to work with or in opposition to nature. This is not to deny that to work against nature might bring adverse consequences, but it is to deny that actions are themselves greatly circumscribed or predetermined by forces outside human control (forces such as the laws of nature, or the so-called 'laws of economics', or metaphysical forces, like those of a god).

Tuan (1972) takes up the Sartrean standpoint on existentialism, which maintains that by virtue of this freedom over his own future man has a freedom relative to other objects. Unlike nature, he is not wholly determined by laws of nature, and he goes beyond what he could do if he were so determined. This fact, coupled with the fact that each human being is a unique individual, means that action in relation to the physical environment, and even what is perceived to *be* the important parts of the physical environment and its significant attributes, will vary from individual to individual, and of course from culture to culture. The emphasis is on the independence and variability of human intentions. This brings us to the extension of existentialism by Husserl, the founder of *phenomenology*. Husserl emphasised the importance of human *intentionality* towards the natural world. If we wish to understand the latter, we can do so only through studying man's intentions towards it and his *consciousness* of it, rather than by trying to study it as some kind of external set of mechanical objects. If man is not conscious of some parts of the natural environment, then to all intents and purposes those parts do not exist - at least they do not exist for man, and any other order of existence is relatively unimportant. Nature, then, has no value (or 'rights') of its own, without reference to man. Indeed, it has no *existence* without him.

35

Differences in emphasis on the importance of man in the man-nature relationship, as epitomised in these two opposing philosophical approaches, will surface repeatedly as we now begin to investigate the origins of modern attitudes identified as 'ecocentric' and 'technocentric'. We shall draw attention to these differences, but would emphasise at the outset that there is no simple correlation between technocentrism and ecocentrism and the free will and determinist philosophies. Both of the former show elements of both philosophies. Thus the ecocentric holds some rather deterministic views, but at the same time he may also emphasise the importance of human individuality and of unique emotional and non-scientific, metaphysical, relationships with, and experiences of, nature. Similarly, the technocentric may wish to emphasise man's freedom to act independently of and over nature, but such freedom may be seen to derive from the close study of the laws of an objectively existing nature and in terms of man's ability to use and work within these laws.

CHAPTER 2

THE ROOTS OF TECHNOLOGICAL ENVIRONMENTALISM

(This chapter is based substantially on material supplied and interpreted by John Perkins)

2.1 TECHNOCENTRISM AND 'CLASSICAL' SCIENCE

Technological environmentalism, or 'technocentrism', as O'Riordan calls it, represents in modern Western societies the *official*, dominant, set of attitudes to the environment. It is the outlook of those groups in society which exercise most power. As described in section 1.5.1, it is characterised by an apparent rationality, a belief in an 'objective' approach, and a conviction that although careful management must be exercised in order to avoid fouling the environmental nest, man is able to manipulate and appropriate nature for his own ends - and is justified in so doing. The high technology and material consumption which is achieved through this appropriation is regarded as the ultimate indicator of social 'progress'. Progress is attainable by knowing and manipulating natural laws and working within the framework of economic laws, *ergo* those who know most about these laws, the objective scientific 'experts', are those in whom trust should be placed when it comes to decision-making about the environment. Because of their relative ignorance the general public are disqualified from participation in this process at any level but the most general.

The roots of this outlook - familiar and almost 'common sense' as it may seem to us - lie in only the most recent past, and in developments which spatially are almost exclusively limited to Western society. Those developments are known as the *scientific revolution* of the 16th to 18th centuries, which was allied to the growth of capitalism. They gave rise to what we are calling 'classical' science, to distinguish it from the 20th-century physical science inspired by Einstein which has undermined many of the bases of 'classical' science. The main component of technocentrism, the official attitude to environment, is classical science and the beliefs and perspectives which it fosters. This science is a major element of the Western cultural filter - it informs, affects and determines our conception of nature at three levels.

One level is *ideological*: science now provides a major element of our basic presuppositions about the world and man's relationship to it, and in coming to do so classical science has displaced alternative ways of understanding the world. It has displaced alternative bodies of natural knowledge, such as myth, folklore and natural magic, and the completeness of this displacement is represented by the pre-eminent position which our society accords to the scientific expert.

A second level is *theoretical*: in its theories science explicitly or implicitly embodies particular concepts of the man-nature relationship - for example in evolutionary theories in biology.

Thirdly, in its *practice* or *methodology*, it involves the scientist in particular methodological positions which describe especially the relationship between himself as the subject - the observer of nature - and the object under observation. Thus in the process - in the very act - of gaining knowledge of nature, the scientist defines his relationship to nature in a particular way.

To understand what these rather abstract statements mean and imply it will be necessary to take an historical approach, which will first involve examining how it was that nature was perceived and understood before the scientific revolution. This examination should make it clear that there is *nothing 'natural' or 'necessary' about our present way of perceiving the universe.* It is not the way that men and women throughout most of history - with as much 'common sense' as ourselves - have chosen to regard it. Therefore, when examining pre-modern perceptions we must not apply modern criteria to them, in some kind of misguided attempt to judge their 'validity' in absolute terms.

In fact, a consideration of our 'modern' perceptions may lead us to the conclusion that it is *we* who take an 'unnatural' and surprising view of nature (as the modern ecocentrics like Capra are trying to tell us). For our view of it, based on classical science, is that it is a sophisticated *machine*, whose behaviour is not random but is a knowable and predictable outcome of its structure. Yet when we look at nature our senses clearly suggest that this is not so - trees, clouds, animals seem to behave rather randomly, and they have qualities, of beauty, colour, life, which we do not generally associate with machines.

The immensity of the change in our perceptions of nature over the past few hundred years can be appreciated by examining, for example, medieval *cosmologies* - theories of the nature, origins and function of the universe. We do this below, and again as a prelude to the examination of ecocentrism in Chapters 3 and 4. From this examination one clear generalisation can be drawn which is crucial to the theme of this book. It seems that before the scientific revolution cultural relationships between man and nature were far *closer and more intimate* than they became as a result of the scientific revolution and the important material developments which called it forth and helped to shape it. Whether the closeness of this relationship meant that man was an exploiter or a steward of nature is arguable - see sections 2.2.2 and 3.1. But the scientific revolution certainly opened up a separation which *enabled* man to exploit nature more effectively than he had previously done. This separation was of a fundamental philosophical kind (see Chapter 5), such that we now find it difficult to conceive of the intimacy of our former relationship. For example, though we still have our occasional superstitions about plants, for us the plant world is hardly any longer 'alive with symbolic meanings' which rest on the 'assumption that man and nature were locked into one interacting world', as was the case even as late as the 18th century (Thomas 1983 p75). By destroying these meanings, natural scientists constructed 'in place of a natural world ... sensitive to man's behaviour ... a detached natural scene to be viewed and studied by the observer from the outside, as if by peering through a window, in the secure knowledge that the objects of contemplation inhabited a separate realm, offering no omens or signs ...' (Thomas p89). To see more what this change in the very definition of the world involved we will now examine aspects of medieval cosmology and cosmography.

2.2.1 Medieval Cosmology

The medieval view of the universe - what intelligent people thought about how it functioned and man's position in it - was governed in its physical aspects by the ideas of Aristotle. These were integrated with evolving Judaeo-Christian ideas over a long period. The integration, which in its philosophical aspects was achieved by St Thomas Aquinas and in science was carried out largely by Albertus Magnus, was a very close one, and was accomplished in the 12th and 13th centuries. There was an almost perfect mapping of Aristotle's physical picture on to Christian theology - onto the Christian moral universe.

The cosmography was a geocentric one. The earth was at the centre of the universe, and, as all evidence suggested, it was solid, stationary, finite and spherical. The stars rotated around and were equidistant from earth (see Figure 4), and they were attached to the inner surface of a rotating sphere which looked like a dome from earth, and marked the edge of the universe. Outside this sphere - beyond the universe's edge - was nothing, or *non-ens* (non-being). This did not mean just empty space, it meant literally that nothing could exist. If one could travel to this boundary and stick one's hand beyond it, the hand would become non-existent. To put it another way, one could not ask questions about this region.

Within the sphere of the fixed stars the universe was divided into two; the celestial and terrestrial zones, with the moon's orbit forming the boundary between them. The behaviour of celestial objects was very predictable, and they moved in circular orbits around the earth at constant speeds. But in the terrestrial region things moved randomly or in straight lines. Terrestrial things were born, died and decayed - they changed. But this did not happen to celestial objects. Non-changingness suggested no need for change because of already-achieved perfection. Thus change meant imperfection. Circular motion suggested perfection - the perfect geometrical figure being a sphere - randomness suggested imperfection. So the celestial zone was one of perfection, the terrestrial zone was one of imperfection. This was an observational *and* a value difference. Observational (empirical) evidence and values were combined.

Between the moon and the sphere of stars were, in order, Mercury,

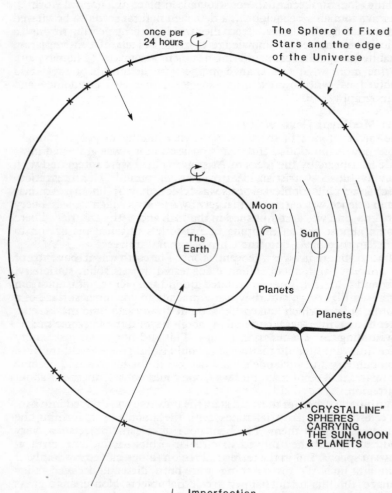

THE CELESTIAL REGION or Superlunary Region { Perfection
Uniform Circular Motion
The fifth Element (Quintessence)

once per 24 hours

The Sphere of Fixed Stars and the edge of the Universe

Moon

Sun

The Earth

Planets

Planets

"CRYSTALLINE" SPHERES CARRYING THE SUN, MOON & PLANETS

THE TERRESTRIAL REGION or Sublunary Region { Imperfection
Rectilinear or Random Motion
Gravity
The Four Elements: Earth, Water, Air, Fire

FIGURE 4   The Medieval Cosmology

Venus, the Sun, Mars, Jupiter and Saturn - each of their orbits was part of a discrete sphere, so the arrangement was of spheres within spheres.

When it came to explaining observed phenomena such as gravity, answers lay in a combination of this observed structure of the universe and the idea that there was *purpose or design* in it. This *teleological* view - one that there was a distant goal toward which all events worked - meant that there was a *final cause* behind everything. Since the cosmology was a Christian one, the Christian God was the final cause. The universe was ruled by principles which helped to achieve God's purposes and he was the final cause of everything. These principles, or physical laws, were a function of God's design and were to be explained through understanding of that design. Such explanations of the physical world which reside in God are known by historians as *physico-theological*. Thus the explanation of gravity was a physico-theological one as follows:

The Earth was made up of four elements; earth, air, fire and water, while the objects in the heavens - this being the region of perfection - must consist of a different, fifth, perfect element - the quintessence. When God created the universe it was symmetrical - the heaviest element (earth) was at the centre, with water surrounding it, then air, then fire. But God then introduced motion into this static universe, spinning the sphere of the stars, and, by friction, this movement was communicated to all the spheres between the stars and moon, and then to the moon. The movement of the last was in turn communicated to the series of earthly elements. They were stirred up and dislodged from their *natural* places (i.e. those designed for them by God). Now the Earth - which had been smooth - had mountains where earth had been thrown up, and valleys and depressions whence it had been dislodged - water had got into these depressions, and fire and air had become mixed in with the other elements.

God, however, intended that the universe should return to its original state. No restoration was needed in the already perfect celestial region, but in the terrestrial zone all the objects were trying to fulfil God's desire. Thus if one held up a stone and then released it, it would fly towards the centre of the earth, where it naturally belonged. It would go in a straight line - the shortest route. This explained gravity, and the well-known acceleration of objects as they descended was also explicable in terms of fulfilling God's purpose. Since the stone wanted to carry out God's desire, and since (as we all know) the nearer one gets to the desired object the faster one goes, the nearer the stone got to its natural place, the faster it went. Thus, behaviour of natural objects was explicable in relation to their position in the universe and as a function of God's desires. Subterranean air, water and fire travelled upwards with force (in volcanoes and springs and geysers) to get to their appointed positions, as did a candle flame point upwards. Water in the region which was appointed for air fell down. The explanations formed part of a remarkably logically consistent, coherent and complete system of theory. Several significant things may be noted about it.

First, this being a physico-theological system, the explanations which scientists gave for physical phenomena followed on from, and were compatible with, theology. The 'paradigm' (see Chapter 5) for science was set by religion. This was so much the case that it was inherent in the academic and power structure of the medieval universities. In them one first studied (from 13 to 19 years of age) in the Arts Faculty. Here, the above cosmology was taught, along with Aristotelian physics and philosophy. One then progressed to the Law or Medicine Faculty, and finished up in the Faculty of Theology. This last was the senior faculty, and it dominated the others in the university hierarchy; and though there was some latitude for varying interpretations, Aristotelian science had in general to conform to Christian theology. So to challenge Aristotle's science would be very difficult - a challenge to this meant also a challenge to the whole theology which encompassed it. The problems of throwing out the cosmology were therefore enormous. Secondly, despite the importance of God in the scheme, the medieval cosmology was *anthropocentric*. It was replete with human views imposed onto nature. Categories used to explain human experience were transferred to explain physical ones; hence the universe was said to be *purposeful*, in which objects *desired* to fit in with God's plan - purpose and desire being human attributes. There were regions of *perfection* and *imperfection*, these being human views entailing *values*. Space had values attributed to it. The celestial region was most valued, but within the terrestrial zone the nearer one went to the earth's centre the greater was the imperfection - with Hell at the centre. And on the earth's surface there were areas more valued than others ('sacred' spaces, like Jerusalem, or the precincts of a cathedral). Scientific knowledge of these spaces, and of behaviour within them, could not be divorced from human values.

Thirdly, and arising from this principle, the world could be seen very much in terms of *analogy* with human experience. *Metaphors* could be used to interpret it, and one must be careful here to recognise that, as Mills (1982) has pointed out, the use of metaphor involved much more than people *likening* X to Y (e.g. nature to a book). To the medieval mind, the metaphor meant that X *was* Y (nature was a book). Nature as a book was indeed one of the major medieval metaphors. It meant that since nature was a result of God's design and desires, one read two books, the Bible and nature, to discover God's purposes and act accordingly. So nature was not there just for the sensuous enjoyment of man; it carried in it instructions and hidden meanings. The industry of the ant or the bee would be an example to man, 'the fly was a reminder of the shortness of life, and the glow-worm of the light of the Holy Spirit' (Thomas 1983 p64, who also reminds us that this metaphor was carried forward until well into the 18th century).

Thus, the charter of medieval symbolism, which flowed from God as the final cause, was that 'There is nothing in visible and corporeal things that does not signify something incorporated and invisible' (the Irish 9th-century philosopher Eriguena, quoted in Mills). In nature one read not only how to be saved but also, because God had created the earth for man (see next section), things of benefit to man on earth. Thus, where a plant, stone or animal resembled in shape or colour or behaviour a human organ or disease, it could be used in healing. This was the doctrine of *signatures*, examples being that milky plants helped new mothers to produce milk, bony plants were good for the bones, summer plants helped to cure summer complaints, (Mills 1982), spotted herbs cured spots, yellow ones healed jaundice and adders tongue was good for snake bites (Thomas 1983).

If the book was a metaphor deriving from human experience, so, too, was the metaphor of the organism. This *organic analogy* was strong from Classical times up to the 18th century, and it has even reappeared today as a major tenet of ecocentrist philosophy (see section 3.1). The medieval version saw the world having feelings and being a huge animal, in which men and women lived, like intestinal parasites. The earth's water and volcanic conduits were its circulatory and digestion systems - volcanoes erupting, for instance, were a process of breaking wind.

Both this analogy and related concepts of pantheism and animism stemmed from the pervasive cosmology which paralleled that of the physical world, whereby the medieval (and later) biological world was organised. That was the cosmology of the Great Chain of Being. In it, the analogy was drawn between all elements in the universe, animate and inanimate, spiritual and material, and the links in a chain. They were joined together in a fixed hierarchy, and were mutually dependent. In the medieval version everything was alive, including stones, earth and 'slime' - given life by the overflowing of the Soul from the perfect God at the head of the hierarchy.

This metaphor and its related ideas tended to place man and nature into an intimate relationship, as we have said of medieval views in general. Furthermore, the relationship implicit in the Chain of Being seems conducive to a degree of humility on the part of man towards nature. It fostered the notion of nature's rights to an existence independent of man since each link of the Chain was equally vital for the whole chain's existence; consequently this concept is regarded here as a forerunner of ecocentric attitudes and is dealt with in section 3.1. However, whether all of the medieval cosmologies placed man and nature on such an equal footing is arguable. Indeed, this matter has been the subject of sharp debate, which we report below. The debate should make us hesitant before jumping to the simplistic conclusion that because there was a profound difference betwen medieval cosmologies and the ideas of the scientific revolution, and because the latter encouraged a domineering attitude by man towards nature, pre-17th-century scientific attitudes towards nature were therefore *not* domineering. One school of thought holds that on the contrary the very Christian theology which subsumed medieval science was tainted by such domineering ideas, that amounted to Christian arrogance.

And of course, whatever the central theological *ideas*, it cannot be denied that in most of medieval Europe the predominant *practice* was to transform nature on an ever-increasing scale, in, for example woodland clearance for open-field agriculture and the iron industry, in fen draining and in building villages and towns (Hoskins 1955). As Thomas (1983 p34) says of the Cartesian ideas which replaced the medieval conception of nature (section 2.3) and aimed explicitly to make men 'lords and possessors of nature', the most powerful argument for their position that animals were machine-like was that it was the best rationalisation of how man *actually treated* animals.

2.2.2 God's Design for Man: to be Nature's Exploiter or Steward?

Medieval cosmologies were infused with ideas which held that the physical world could be explained by reference to God's design for it. Glacken believes that this notion was extant up to the time of Lyell and Darwin despite the scientific revolution which we describe below. Indeed, though the tendency of the latter was to substitute for explanations involving divine purpose theories which were a product of the universe's innate mathematical character, this does not mean that the chief characters of the scientific revolution thereby rejected the doctrine of God as the *final* cause. Glacken holds that Galileo, Kepler, Newton and Boyle did not deny its validity, while Bacon did not 'object' to it. And Leibnitz felt that the very existence of God was proven by cosmic harmony and order, which derived from the one unifying principle (see 3.1). The argument is, then, that medieval cosmologies were essentially organic, holistic and harmonious, tending to foster egalitarian man-environment attitudes. Interrelationships were important; these and harmonies and adaptations to the environment stemmed from the work and design of God. Such thinking is found in John Ray's (1691) *The Wisdom of God Manifested in the Works of Creation*, where the author divorces himself from the emerging scientific view of nature as a machine and separate from man.

This, however, is but one interpretation of physico-theology. Lovejoy (1974) is one of several who take a different view in a continuing debate about whether Christian teachings inspire exploitative or harmonious attitudes to nature. He thinks that physico-theological ideas came to rest on the assumption that all other elements of nature were created *for man's sake*, and therefore put man and nature into an *unequal* relationship:

> *Tout est créé pour l'homme* is at once the tacit premise and the triumphant conclusion of that long series of teleological arguments which ... is one of the most curious moments of human imbecility (Lovejoy 1974 p186).

Lovejoy goes on to argue that this arrogant attitude was carried into classical science - he thus sees here a continuity of thought. There may be some inconsistency in his view, for Lovejoy also sees the Great Chain of Being concept - a physico-theological one - as conducive to human *humility* towards nature (see 3.1). But then, the Chain of Being, too, can be regarded as ambiguous in meaning. For it could equally be deduced from it that man's position in the Chain is *superior* to that of the rest of nature and therefore this gives him the right to dominate and exploit nature (Thomas 1983 p124 and Glacken 1967 pp482,600).

White (1967) certainly believes that this sentiment constituted the legacy of Judaeo-Christian thought, and he argues for its appearance in early Middle Ages, and its continuity up to the present. Allied with technological and scientific developments, orthodox Christianity has produced arrogant exploitation of nature, and a contemporary ecological crisis. White's thesis is that the West's successful science and technology developed between the 8th and 12th centuries - it is much older than the scientific revolution though it was not until about 1850 - following the democratic revolutions - that science and technology were combined to produce truly immense powers to change nature. The early development, however, was paralleled by the development of exploitative attitudes to nature which seemed to be 'in harmony with larger intellectual patterns', namely the victory of Christianity over paganism. This destroyed the animistic beliefs whereby men thought twice before they plundered and destroyed natural objects. It substituted instead a faith in perpetual progress, a belief that God designed nature for man's benefit and rule, and that action, not contemplation, was the correct Christian behaviour. Science formed an extension of theology (for to know God you had to find out how his creation worked), and technology provided the active means to carry out God's will. Because today's attitudes are essentially inherited from Christianity, then it 'bears the burden of guilt' for contemporary ecological disruption.

The linchpin of White's thesis, which Thomas and Nash (1974) also seem substantially to accept, is the Book of Genesis, with its commands to man to multiply himself and spread over the earth and make his dominion over it. And while Adam was placed in the Garden of Eden as its caretaker - to 'till and keep it' (Gen 2:15), after the flood 'Every moving thing that lives shall be food for you; and as I gave you the green plants I give you everything' (Gen 9:3).

These passages were selected by Glacken, but by reference to Job, and St Paul's Epistles to the Romans, this writer also shows that the Judaeo-Christian legacy can be interpreted in exactly the *reverse way*. For these Books of the Bible show that though man is 'set apart from all other forms of life and matter because God has willed this role for him; he is ... set here as a steward, responsible to his Creator for all he does with the world over which he is given dominion' (Glacken 1967 p152). Doughty (1981), too, believes that there is a theological perspective in Christian thought which is sympathetic to the environment (but does not necessarily involve turning to a radical like St Francis of Assisi as White recommends). He regards White's argument as 'partisan and overgeneralised', and he cites scholars like Santmire and Gowan, who maintain that Genesis characterises man as a life tenant of nature, not a freeholder. He also quotes Bishop Montefiori's belief that *only* by living by the gospels will man be saved from his environmental predicament, for they contain moral obligations to nature and the comprehension of limits imposed by natural law. Attfield (1983), too, maintains that 'there is much more evidence than is usually acknowledged for other, more beneficent Christian attitudes to the environment and to nonhuman nature'. Christian teachings about nature have been diverse and full of contradictions and difficulties, but they have, he says, not typically been exploitative. He asserts that Psalms 104 and 148 cast doubt on Passmore's (1974) claim that the Old Testament suggests that everything exists to serve humanity. For these Psalms admire God's handiwork, express his care for a variety of creatures, and suggest that man's domination of nature means ruling it in a way consistent with being responsible to God for his realm. Similarly, the New Testament, in Matthew's and Luke's gospels, also bespeaks God's care for animals

45

and plants. As to White's thesis that Western science was replete with exploitative attitudes to nature, Attfield points out that the earliest adherents of modern science, such as John Evelyn, were far from believing that they could treat nature as they pleased, while even Descartes did not suppose that animals had no feelings (see Cottingham 1978). Attfield concludes that adverse interpretations of Christian attitudes to nature are at times derived by the selective use and exaggeration of evidence.

Certainly White's argument, compelling though it might appear at first, does not really withstand close scrutiny. It is vague about dates, using them with abandon and imprecision. Its language is speculative, establishing no causal connections. It is technologically determinist, maintaining that technological changes *cause* dominant ideologies to change, and it does not consider closely the relationship between the ideas it discusses and economic and social changes other than technological ones. How Christian ideas (of dominance over nature) appear to have survived the major upheavals of the scientific and industrial revolutions with so little apparent modification is not made clear.

Instead the truth seems to be that *both* ideas - man as dominant or as a steward over nature - can be found in Christianity. The question arises as to why the former seems to have been *selected* as the dominant (technocentric) ideology with the advent of the scientific revolution and the rise of capitalism. Glacken's view is the one we incline to. He says (p494):

> In the period roughly from the end of the 15th century until the end of the 17th century one sees ideas of man as a controller of nature beginning to crystallise, along more modern lines. It is in the thought of this period (not the commands of God in Genesis to have dominion over nature) ... that there begins a unique formulation of Western thought, marking itself off from the other great traditions, such as the Indian and the Chinese ... The awareness of man's power increases greatly in the 18th century ... It increases even more dramatically in the 19th century, while in the 20th century Western man has attained a breathtaking anthropocentrism, based on his power over nature, unmatched by anything in the past.

We now turn to the thoughts of this period for an understanding of this 'breathtaking anthropocentrism' which is one of the hallmarks of technocentrism.

2.3 THE SCIENTIFIC REVOLUTION AND NATURE AS A MACHINE

The period known as the 'scientific revolution' took up about 150 years, from the time of Copernicus, who published, in 1543, *On the Revolutions of the Heavenly Orbs*, to the end of the 17th century, with Isaac Newton's *Mathematical Principles of Natural Philosophy* (1687) and his *Optics* (1702). During this period the principles of 'classical' science were being established in contradistinction to the medieval cosmology described in section 2.2.1. These principles are sometimes known as the 'Newtonian paradigm' - that is the model, derived from Newton's work, of what questions it was legitimate for science to ask of nature, how they should be asked and what constituted valid answers (see Chapter 5). Despite Einstein, quantum theory, and all the other advances associated with 20th-century physics, and despite the advocacy of 'holistic' and 'organic' systems approaches in science (see Chapter 4) it is nonetheless true that the Newtonian paradigm still constitutes the foundation of the modern popular conception of science, and it is this conception which is also the basis for technocentrism.

The scientific revolution took up a long period because, as we have suggested in 2.2.1, it involved a fundamental challenge to the medieval cosmology based on divine purpose. Copernicus suggested a simple revision to the cosmography whereby the positions of the sun and earth would be reversed, but the implications of this were so enormous that it needed 150 years to construct the new cosmology which was required. It should be remembered that the new ideas served not only as an intellectual challenge to the established science. They also ate away at the theology from which it stemmed - which in turn supported a particular social structure. And were it not for the social and economic challenges to that structure which, too, were being made, it is doubtful whether the intellectual ideas represented by the Newtonian paradigm would have triumphed in the 18th and 19th centuries as they did. We should also bear in mind that although through 300 years of socialisation the ideas of the Newtonian paradigm constitute 'common sense' to us, they are not that 'sensible' - that is, they do *not* accord with what the most immediate evidence of our senses tells us about the world. When we look around us, it is apparent that neither we nor the clouds nor the other objects we see are speeding along at about 1000mph. And it is equally obvious that a 'solid' object like a stone *is* solid, and not composed partly of empty space. Classical Newtonian science tells us that we are wrong, and that what we think we see is unreal.

Copernicus' proposal for a heliocentric solar system opened up, then, an enormous can of worms, and left science to explain difficult problems which had previously been easy to account for. If, for example, earth moved through the universe, space could not be divided up into perfect and imperfect regions. Also, night and day could not be explained in terms of the movement of the stars and sun around the earth, therefore the latter had to be conceived of as rotating. But this then raised questions of why we do not encounter the 1000mph wind as we rush around on our axis, and why objects which we throw vertically upwards also land vertically instead of some distance away. Neither could this vertical fall be explained as before as a function of a body's position in space, and its desire to reach its natural position. And the removal of the distinction between perfect and imperfect space also meant revising theories of matter, whereby the four elements had been spatially distinguished from a quintessence. The new theories would have to be based on *universally-applicable* principles.

Johannes Kepler (1571-1630) sought just these universal principles when he tried to explain the movements of the planets. He argued that the sun was the cause; both it and the planets functioned as magnets, and in rotating the sun pushed the planets around with it. So the sun was the driving force to a kind of *machine*:

> I am much occupied with the investigation of the physical causes. My aim in this is to show that the celestial machine is to be likened not to a divine organism but rather to a clockwork..., insofar as nearly all the manifold movements are carried out by means of a single, quite simple magnetic force, as in the case of a clockwork all motions (are caused) by a simple weight. Moreover I show how this physical conception is to be presented through calculation and geometry (Kepler, letter to Herwart von Hohenburg, 1605, quoted by Holton (1956)).

47

Thus Kepler was using a new metaphor for nature - the organic analogy was to be replaced by that of a clock. This was the metaphor used by all of the prime movers of the scientific revolution, and the *mechanistic conception of nature* is a major component of the classical scientific (technocentric) view. Kepler was religious, and he believed he could understand nature and, through it, God's intention, via mathematics and geometry. In fact God's mind was geometrical and geometry was God:

> Why waste words? Geometry existed before the Creation, is co-eternal with the mind of God, *is God himself* (what exists in God that is not God himself?); geometry provided God with a model for the creation and was implanted into man, together with God's own likeness - and not merely conveyed to his mind through the eyes (Kepler, *Harmonici Mundi* (1619), quoted by Koestler (1964)).

This was, therefore, a conception of God as a Creator and nothing more. He was not ever-present and did not intervene in the workings of his creation. Rather, he was an *engineer*, using geometry to make a plan from which he had constructed a machine. He set the machine going and then left it. This was an essentially *deterministic* view of nature (see Chapter 4, and section 5.2.1). For the way a machine works is that its structure and past configuration determine its present behaviour. As Capra (1982 p52) puts it:

> In this way [the explanations afforded by the Newtonian paradigm] the whole universe was set in motion and it has continued to run ever since, like a machine governed by immutable laws. The mechanistic view of nature is thus closely related to a rigorous determinism, with the giant cosmic machine completely causal and determinate. All that happened had a definite effect, and the future of any part of the system would - in principle - be predicted with absolute certainty.

(Note that the past now determined the present rather than, as in the medieval physico-theological view of a final cause, a future goal determining present behaviour.)

Kepler's view was developed by Galileo Galilei (1564-1642). He believed that the 'book of nature' was written in the language of mathematics and therefore had to be 'read' and understood via mathematics. The physical problems of motion were to be treated as geometrical problems, and objects like falling stones as geometrical entities. This followed from the necessary structure of God's creation, for God had structured the universe according to mathematical principles. Consequently the way to understand it was through deduction from these principles in conjunction with observation and experimentation.

So nature was a machine whose workings were governed by and predicted through mathematical laws. The logical outcome of this mechanisation and mathematisation of nature was that what we see when we look at nature is not what is real (for we do not see a machine). The true reality of nature is mathematical and regular. From this followed what has become a basic tenet of classical scientific philosophy (see section 2.5): *what is truly real is mathematical and measurable, and what cannot be measured cannot have true existence.* In essence, it was Galileo who first distinguished between what was measurable and therefore objectively ascertainable - being the same for everyone - and what was not measurable and therefore varied from person (subject) to person (subject):

> Now I say that whenever I conceive any material or corporeal substance, I immediately feel the need to think of it as bounded, and as having this or that shape; as being large or small in relation to other things, and in some specific place at any given time; as being in motion or at rest; as touching or not touching some other body; and as being one in number or few, or many. From these conditions I cannot separate such a substance by any stretch of my imagination. But that it must be white or red, bitter or sweet, noisy or silent, and of sweet or foul odour, my mind does not feel compelled to bring in as necessary accompaniments. Without the senses as our guides, reason or imagination unaided would probably never arrive at qualities like these. Hence I think that tastes, odours, colours, and so on are no more than mere names so far as the object in which we place them is concerned, and that they reside only in the consciousness. Hence if the living creatures were removed, all these qualities would be wiped away and annihilated (Galileo, *The Assayer*, 1623).

Using the technique of 'systematic doubt', Galileo here argued that whatever else he was unsure of he could not doubt that objects had shape, size, motion and quantity - he could not, in other words, imagine objects without these (measurable) properties. But he could conceive of bodies with or without one or other sort of smell or taste or colour. So these properties were variable and existed only in human consciousness. The properties were not totally necessary or inherent qualities of the objects, as witness the fact that what they were (and whether they existed at all) was a matter for different individuals to judge differently upon. Thus if humans were removed from the scene, objects would still have shape, size, position, motion and quantity - these were what really existed 'out there', and they were *primary qualities*. But the rest of the qualities would not continue to exist without the existence of humans. They were not really 'out there', and were therefore *secondary qualities*. It followed that if one wanted to understand what *really* existed, one would have to try to cut out the influence of the secondary qualities. One had to cover up one's personality and the influence of the (subjectively-ascertainable) secondary qualities and restrict one's attention to the (objectively-calculable) primary qualities.

Primary qualities, then, can be measured; secondary qualities cannot. Primary qualities have objective existence; secondary qualities are subjective. Primary qualities are not challengeable; secondary qualities, by contrast, are not 'true' or 'real'; they are just a matter of opinion. Therefore the truth of scientific theory has nothing to do with how soundly or cleverly that theory is argued, but with how *objective* is the knowledge from which it stems: *objective knowledge is 'true', and correct, while subjective knowledge is not.* It is in this distinction that classical science still has its powerful appeal as being in some way *independent*. For it can give 'objective' knowledge of nature. Being objective rather than subjective, this knowledge must be free from the sectional interests of any particular person or group - it can therefore be *trusted*.

With this distinction between what is really 'out there' - objects with primary qualities only; grey and moving mechanistically in space - and what is in the minds of men - subjective qualities like beauty and colour - arises also a fundamental distinction between man and nature. An enormous gulf, in fact, opens up between the two.

The gulf was most starkly revealed by René Descartes (1596-1650), the most radical of the thinkers of the scientific revolution, and one of the creators of modern philosophy. Developing on, and going further than, Kepler and Galileo, he argued that matter was nothing more than extension in space. It was geometry. It did not consist of sensible qualities like hardness, weight or colour, but was extension in breadth, length and depth. What was mathematical was real and what was real was mathematical. The implications of this were that the universe must be infinite (because the space of a geometer was infinite) and that since matter was extension in space the universe must be full - there could not be empty space. The only way to have a full universe which also contained motion was to have matter divided into particles. For Descartes, matter was infinitely divisible, and the universe contained nothing except particles with size, shape, weight, motion and position. Explanations had to be in terms of matter - particles - in motion. No occult action, in terms of sympathy or desire, was possible.

Descartes extended the concepts of nature as a machine and the unreality of that which was not measurable. He viewed animals, and the human body too, as machines. They were automata, and their workings could be fully known by reducing them to matters of physics and chemistry, which in turn would be understood in terms of mathematics. This is a *reductionist* view, whereby, through *analysis*, everything can be reduced to the same basic quantities and qualities, all of which are measurable and expressed in terms of *universally*-applicable principles. It holds that there is no vital principle necessary to life - life differs from non-life only in its degree of complexity. Reductionism and universalism abound in the classical scientific view, which is brought up to date, for example, in the Royal Society standards for 1975. These choose seven physical qualities as 'dimensionally independent base qualities' (length, mass, time, etc.). 'All other physical qualities are regarded as derived from these basic qualities' - quoted by Hales (1982 p124-5) who goes on to say 'These plus the ten base digits 0-9 ... are the sum total of the legitimate language of observation in hard physical science, and the stuff of all explanation'.

Capra's and Skolimowski's ecocentrist critiques of classical science are aimed particularly at its mechanistic and reductionist tendencies. These tendencies made it possible for scientists to argue that consciousness could be explained mechanically and that a person's entire psychic life 'was the product of his physical organisation. What Descartes said of animals would one day be said of man' (Thomas 1983 p33). Reductionism also implies that because mathematics is the language of science - describing the laws governing behaviour of its ultimate particles - the measure of what is scientific is provided by how *mathematical* it is. Hence modern debates in the social and economic 'sciences' about their development *as* sciences hinge in part on the degree of quantification they are undergoing.

In view of the reductionist principle, the question arose whether, and in what way, man was at all distinguishable from the rest of nature. Descartes solved this problem through, again, the process of systematic doubt:

> But then, immediately as I strove to think of everything as false, I realised that, in the very act of thinking everything false, I was aware of myself as something real; and observing that the truth: *I think, therefore I am*, was so firm and so assured that the most extravagant arguments of the sceptics were incapable of shaking it, I concluded that I might have no scruple in taking it as that first principle of philosophy for which I was looking.
>
> My next step was to examine attentively what I was, and here I saw that, although I could pretend that I had no body, and that there was neither world nor place in which I existed, I could by no means pretend that I myself was non-existent; on the contrary, from the mere fact that I could think of doubting the truth of other things, it followed quite clearly and evidently that I existed; whereas I should have had no reason to believe in my existence, had I but ceased to think for a moment, even if everything I ever imagined had been true. I concluded that I was a substance whose whole essence or nature consists in thinking, and whose existence depends neither on its location in space nor on any material thing. Thus the self, or rather the soul, by which I am what I am, is entirely distinct from the body, is indeed easier to know than the body, and would not cease to be what it is, even if there were no body. (Descartes, *Discourse on Method*, 1637)

Descartes' reasoning was that the very act of doubting was a process of thinking, that the one thing that could not be doubted was that *he* was thinking and that this thinking was the act which established the fact of human existence. Humans were therefore definable as no more nor less than thinking beings. What separated them from the rest of nature, and from their own bodies, was this thought process. Whereas the body could be reduced, or analysed, into its component parts, the mind could not. Descartes thereby had introduced a most fundamental dualism in modern thought - that between mind and matter. Whereas matter was composed of primary, objectively-knowable, qualities, the mind was subjective and attributed secondary qualities to nature. Thus the *Cartesian dualism* involved mind and matter, subject and object, and it had a profound implication for the man-nature relationship because nature became composed of *objects metaphysically separated from man*. These objects had primary qualities and no others. They were reducible to atoms, whose unthinking, machine-like behaviour was universally the same and explicable in terms of mathematical laws. Man, by contrast, was defined as a rational thinking being - the subject who observed the object and could impart, in his mind, secondary qualities to nature.

It was this dualism, rather than any specifically Christian doctrine, which paved the way for a man-nature separation in which the former was conceived of as superior to the latter, though Descartes himself did not draw from it the conclusion that nature was created for man alone (Thomas 1983 p167, Lovejoy 1974 p188).

In terms of the 17th century, Descartes took his ideas beyond the scope of the thinking of most of his contemporaries. Indeed Newton formulated his own theories of motion, gravitation, and light as a critique of Descartes' explanations. Nevertheless, Descartes' mathematical analytical method and his metaphysics - in particular the dualism of mind and matter, subject and object, and nature-as-machine - were shared; if not so much by Newton, then by the scientists in France and England in the 18th century who established the Newtonian paradigm as scientific orthodoxy. This classical Newtonian science was analytical, experimental and reductionist, seeking understanding by taking the machine of nature to pieces to see how it worked. Mathematics was to be its language; it sought, as Newton had indicated in the title of his greatest work, the mathematical principles of natural philosophy. It set out to explain the phenomena of nature in terms of nothing more than matter, composed of indivisible particles called atoms, in motion in infinite geometrical space under the influence of forces that were measurable. Unlike medieval Aristotelean science with its universe divided into two distinct regions, these explanatory principles were to apply *throughout the universe*. Thus, when in the story Newton was hit by the falling apple, he did not conclude that he had merely learnt about the behaviour of apples in orchards, but that he had understood how the moon moved, because his science was *universal*. The same principles, explanations and causes were seen to operate on earth, on the moon, everywhere: Newton formulated not the law of gravity, but the law of *universal* gravitation.

Embodying the ideas discussed in this section, classical Newtonian science encapsulated the principles of universalism, analysis, objectivity and reductionism and established the defining characteristics of science (section 2.5) until at least the early 20th century. For the foundations of technocentric attitudes about the purposes of that science and its methods, and to the social role of the scientists, we need to turn next to Francis Bacon (1561-1626). But before we do so there is an important omission to rectify.

So far we have presented the scientific revolution as a transition from medieval Aristotelian science to classical Newtonian science, from the universe conceived as a purposeful, divine organism to nature understood as a mathematical machine. This view is over-simplified and before we go on to discuss some of the consequences and implications of the establishment of classical science we must attempt a more complete picture.

Historians of science have recently directed much research onto a third system of natural knowledge, known as *natural magic*. This exerted a powerful hold on European thought, in the form of alchemy and astrology, during the Renaissance and the period of the scientific revolution. It was at least as widespread and as influential as Aristotelian science, and we must therefore see the scientific revolution as a three-cornered conflict from which classical Newtonian science eventually emerged as the new orthodoxy.

The basic tenets of natural magic were, again, that the universe is an organism, fully alive and active. It was permeated with influences, forces and correspondences that linked everything in nature, man included, to everything else, forming a multi-dimensional network that was not only material but also mystical and spiritual. Thus, everything that was found in the small universe that was man (the *microcosm*) - the organs of his body, his intellect, senses, spirit and emotions - was linked with, and corresponded to, the various parts of the larger universe of nature (the *macrocosm*). For example, man's heart corresponded to the sun; the sun was the heart of the universe; the sun was the symbol of the heart and vice-versa: in a sense the heart and the sun were the same. The microcosm and macrocosm were images and symbols of each other - two levels of the same reality. Hence one could study man by studying the universe and the understanding of man would be at the same time an understanding of the universe. Furthermore, the violation of nature (say, through mining) constituted the violation of a human (female) body. Consequently natural magic was opposed to any suggestion that a distinction had to be drawn between man and nature, as subject and object, as a precondition for understanding nature: the contrary was the case. The natural magician had to recognise that he was inextricably part of the nature he was studying. His knowledge was therefore not (and he would not have wished to claim that it was) objective and impersonal: it was subjective and personal. Furthermore, reason had to be inadequate to the task of understanding a universe permeated with occult forces and symbolic relationships, only the non-logical faculties of intuition and empathy were capable of doing so. These faculties were brought into play when the natural magician immersed himself in nature and opened himself to the interplay of her forces and influences, and he did this by using and manipulating those forces, that is, by experimentation. Natural magic, alchemy in particular, was profoundly experimental, but in all other aspects it was deeply opposed to the classical science emerging alongside it in the 17th century. It rejected the basic distinction that Galileo and Descartes were drawing between subject and object. Measurement, so central to Galileo and Descartes, it rejected as incapable of dealing with a world whose fundamental elements - influences, correspondences, symbols - were immeasurable and immaterial; nature-as-machine could not do justice to its richer vision of the world.

It was as much in opposition to natural magic as to Aristotelian science that Galileo drew the distinction between primary and secondary qualities and that Descartes pronounced his strictures against explanations that relied on the occult qualities of sympathy, desire and action-at-a-distance. The wide differences between natural magic and classical science force us to recognise once again the radical nature of the latter, and to say that its fundamental assumptions about the world and man's relationship to it are by no means natural or common sense. The existence of a powerful rival in the form of natural magic indicates that there was nothing natural or automatic about the eventual triumph of classical science. To understand fully the reasons for it we would have to investigate the historical forces and circumstances through which classical science rather than, for example, natural magic, came to constitute scientific orthodoxy.

2.4 THE BACONIAN CREED AND ITS HIGH PRIESTS

Bacon was the first figure in the scientific revolution to draw out the full implications for the man-nature relationship of the emerging principles of the 'new science'. As we have noted, medieval cosmology asserted the unity of man and nature; Cartesian dualism effectively separated the two. But it was Bacon who asserted the creed that *scientific knowledge equals power over nature*.

Bacon and Descartes represent two fundamental strands in classical science. The former may be viewed as pragmatic and practical while the latter was more abstract and speculative. Nonetheless, Bacon grappled with the same problem as Descartes, which was the separateness of man from his object of study, nature. The subjectivity of the former constituted an impediment to his gaining knowledge of the latter. Man, the subject, was forced to interpret nature through human feelings and experiences; he had selfish interests, assumptions and presuppositions. All these militated against his being objective, while the words of his language were imprecise. Thus the unfettered exercise of the human mind through reasoning constituted an inadequate scientific method. Reason had to be controlled by the evidence of the senses. Descartes started with reason, by formulating hypotheses based on assumptions about the laws governing relationships between objects, and he argued that from them the expected behaviour of the objects could be deduced. If, then, observed matched expected behaviour, the assumptions were deemed correct. His method was thus *deductive*. But Bacon's method was the reverse; it was *inductive*. It held that the scientist should *first* make many observations of nature, and from them it would then be possible to draw out, or induce, laws governing their relationships. Having been formulated through observation, hypotheses could then be tested and verified by the collection of more data, and once verified these hypotheses would gain the status of laws of nature. Further observation and experiment would lead to the formulation of laws that were wider and more general in scope, with the distant goal of a single law that would embrace all the phenomena in the universe. Bacon's scientific method can be visualised as pyramidal - founded on a broad base of empirical (based on observation and/or experiment) knowledge, and building on to it laws of nature of increasing generality; the whole pyramid crowned with the universal law that will explain everything. In practice, classical science followed Newton in mixing and combining deductive and inductive approaches. But the importance of Bacon's induction lay in its implications that 'truth is the daughter of time' and that science is progressive.

For the collection of empirical knowledge was a long process. Inductive science was based on the steady accumulation of observed data and a process of experimentation. New knowledge was built upon old knowledge. Thus scientific knowledge increased with time; it could not be encompassed by any one individual or age. It was a communal activity, in which the community of scholars was working towards a future goal - the establishment of truth (although the ultimate truth could never be known, for this was the exclusive province of God).

Such activity could not be accomplished within the existing scientific institutions, for the medieval tradition was too strong in them, whereby argument rather than objectively-assessed facts was the hallmark of valid knowledge. Hence Bacon advocated a new start for science, in separate research institutions that would be state-supported.

Bacon went further and argued not only that science should receive this recognition and support but that it was central to human endeavour because of its goals and the motives associated with doing science. Thus the true understanding of nature that only the inductive method could achieve was first and foremost the means of glorifying God, who had created nature.

The case for state support required justification, and Bacon embodied this in his definition of the purpose of scientific activity. It was, first, done to glorify God. Second, it aimed to relieve man's estate. It was a *philanthropic* activity in which the scientist should assume his moral duty of improving man's material lot. This improvement should come about by understanding how the machine of nature worked. Understanding the laws of nature was the first stage in *using* those laws for man's benefit. So science's purpose was to 'command nature in action' - 'The end of our foundation is the knowledge of causes and secret motions of things and the enlarging of the bounds of the human empire to the effecting of all things possible' (Bacon, *New Atlantis*).

And there was no compunction about doing this. 'For the whole world works together in the service of man; and there is nothing from which he does not derive use and fruit ... insomuch that all things seem to be going about man's business and not their own' (Bacon, *De Sapientiae Veterum*).

Science was therefore progressive in two senses. First, in that it built upon the secure basis of facts, advancing from them towards greater and greater truth. Second, this steady march to truth would be the way to obtain a progressive improvement in the material circumstances of humanity. Thus for Bacon science was *equatable with human progress*, a powerful argument for supporting a community of scientific scholars - of professional scientists.

With the professionalisation and socialisation of science its role became increasingly identified with this utilitarian humanist objective, and less tied to the medieval one of knowledge of God. In fact many ecocentrics (e.g. Goldsmith 1975A) think that science itself has now become a religion (see section 2.5). If so, scientists are its priests, and this image was also a feature of the Baconian view.

'Democratic', 'compassionate', 'humble', 'radical', 'socially aware', 'philanthropic', 'honest', 'unselfish', 'serene', 'noble', 'dedicated', 'priestly', 'good', 'cosmopolitan', 'apolitical' - all these epithets are suggested by Bacon's descriptions of scientists (Prior 1954).

In his essay *New Atlantis* he portrayed a mythical island where a community of scientists were accorded a high position in society. Their leader bore the stamp of dignity - he was a peaceful and serene man, pitying and democratic. The community was dedicated to gaining knowledge not for profit, fame or power, but for the benefit of all. The members accepted no authority except that of the scientific method. They bore no allegiance to any particular social group, and withheld their knowledge from the state if they felt it undesirable for the state to have it. So objective knowledge was produced and protected by a group dedicated to humanity as a whole. Therefore, anything leading to the development of science was, by implication, good. And also by implication it was scientists who should have the major decision making roles.

This view of the professional scientist: objective, undogmatic, internationalist and committed to improving man's lot, can easily be equated with that of a priest, replacing the established religious priesthood. He was working, it seemed, for a *univeral* good, for the interests of science are universal. This was science's self-justification, and it was a powerful one which would appeal to any group wanting to legitimate its own activities. For if the *group's* interest could be equated with the *general* interest by identifying itself with science, then the group's appeals for support, recognition or power would be more acceptable, or less easy to resist.

This kind of reasoning underlies the explanation of why and how classical science has become the dominant ideology in the 250 years or so following the scientific revolution. Its success has been phenomenal - its perspective has come to be equated with 'natural', 'normal' vision - and it has become the pursuit of most European intellectuals, in place of 'natural magic' such as alchemy and astrology, which were but two of a number of perfectly respectable non-rational approaches to nature in the 17th century. Science has also become the arbiter of most environmental and many social issues - it is appealed to as a source of objective truth on which to base decisions.

A popular modern technocentric view of the reasons for this is that 'it works - one only has to look at the things science has made possible'. These include satellite communication, organ transplants, 'bombs that can destroy entire cities. These things suggest that, for good or ill, scientists get their sums right. Science has much in common with common sense...' (*The Economist* 1981). As has been shown, this last sentence is quite inaccurate, and we should be sceptical about the rest of the statement as well. For it suggests that science's progress can be viewed apart from its *social* context, as somehow a function of its ability to achieve technically workable and sophisticated designs. First, however, one needs a society where this ability is prized as part of a materialist philosophy, to such an extent that the ills are acceptable alongside the good.

Science is a *social* activity, the scientist is a professional, and professionalisation was a *social* process. It was a process whereby science appeared outside the medieval university, in the context of scientific societies such as the Royal Society and the Paris Academy of Science, both founded in the 1660s. These were times of immense change and upheaval in Europe, with the breakdown of Christian certainties and Catholic consensus. Feudalism was collapsing, and capitalism was emerging within the feudal society. The class structure of society was changing. As Hales (1982) puts it, merchants, 'improving' landlords, administrators and others with an interest in the development of manufactures and trade organisations were asserting their growing power against that of 'court-favoured magnates and traditionalist aristocratic landlords. Science became part of the cultural identity of those who endorsed and propagated mobile and calculating anti-traditionalist values...'. In other words, contrary to the Baconian model, science was not neutral and apart from the interests of specific social groups. And religion and politics were ruled out of order in the Royal Society 'not because of the "neutrality" of those who supported the Baconian creed of material power, but because it was politic to suppress views on such matters at a time when power broking was a very active process' (Hales p87). The rise of classical science was not achieved without its becoming allied with rising social groups, and useful (because of its Baconian image) to those groups. It was new, *revolutionary* and, as we have seen, associated with social progress. Later, once the new social order was established, another image drawn from science was used to *maintain* that order - this was the image of the 'naturalness' of order, discipline, hierarchy, competition and struggle (see Thackray 1974). This theme, of the historical association of classical science with particular social groups, is important in deciphering the messages which science gives us about nature and society, and it is therefore taken up in more detail in Chapter 5.

But if social factors influenced scientific development, it is also true that the reverse was the case, and that scientific ideas influenced ideas on social development. In fact, science played a central role in the emergence of a new secular view of man and society in the 18th century Enlightenment. Fundamental to that view is the concept of *progress*.

Writers in the 18th century from Voltaire (1694-1778) to Condorcet (1743-1794) argued that since the end of the Middle Ages society had been changing at an ever-faster rate and that this change was for the better. They held that this had not been true of the Middle Ages because the superstitions of the Bible and the false science of Aristotle had kept mankind in darkness and ignorance, oppressed and enslaved. But now the new science of Copernicus, Galileo, Descartes, and above all Newton, had banished these errors and superstitions, discovered the truth and brought mankind out of the Dark Ages. The success of Newton's science had shown that mankind must no longer seek knowledge in the old authorities - Aristotle and the Scriptures - but must seek it in nature herself by using reason. Man must not accept what he was told, but must free himself, by the independent use of his reason, from the dead hand of the past.

For the 18th century, Newton's great triumph had been in showing that the enormous complexity of the universe was intelligible and that its true reality lay not in the chaos of its surface appearances but in the harmony and simplicity of its inner nature, which could be discovered by reason and experiment and expressed in mathematical laws. When the Enlightenment *philosophe* looked at his own society he found injustice and inhumanity, oppression and slavery, but he had learnt from science that the appearance of things is not the way they are. Reason and experiment would enable him to penetrate beneath this surface to discover the natural laws of a harmonious social life. This was to be the task of a new science of society and a new science of man, modelled on Newtonian science. Once these laws had been discovered it would then be possible for rational men (and the 18th century accepted Descartes' demonstration that man is a rational being) to act to modify society in such a way as to bring present social arrangements into closer conformity with the natural laws governing social behaviour. And this would be an improvement - progress towards a just and more humane society. Reason and science were thus the instruments by which error and superstition, tyranny and oppression were to be attacked and removed, and the means by which mankind was to take control of his own destiny and to set out to create a better future. The Enlightenment thereby extended Francis Bacon's argument: science was to be not just the means of improving man's material circumstances, but, more than that, the means of commanding *human* nature in action so as to improve his social and moral condition. Just as, in consequence, Bacon had assigned to the scientist a superior social role and position, so the Enlightenment gave an elite composed of rationalist intellectuals (*philosophes*) and social scientists (though the term was not yet invented) the major role in the research, education and action that would bring about progress.

The improvement of man's moral condition, the perfectibility of humanity itself, became possible through the adoption of John Locke's theory of knowledge, which owed much to the philosophy of Francis Bacon and the science of the 17th century. In his *Essay Concerning Human Understanding* (1690) Locke argued that at birth the human mind was a *tabula rasa*, and that knowledge consisted of the accumulation on that clean slate of impressions transmitted from the environment by the senses. It followed that all men were in this respect equal at birth and developed entirely by interaction with their environment. Man was his experience. Changing that experience, by instituting the rational natural laws of social organisation and by rational scientific education, would change man for the better. The potential for progress was infinite. As Condorcet, mathematician and perpetual secretary of the Paris Academy of Sciences, said in his *Sketch for a Historical Picture of the Progress of the Human Mind:* (1794)

> Nature has set no term to the perfection of human faculties, the perfectibility of man is truly infinite; and... the progress of this perfectibility, from now on independent of any power that might wish to halt it, has no other limit than the duration of the globe upon which nature has cast us.

Condorcet's *Sketch* is the most powerful statement of the new 18th-century view of man and society. It is one in which science is the instrument of progress towards a free, just society in which all men and women, equal in being distinguished from nature by their power of reason and the ability to manipulate nature, have equal claims on natural justice, rights and law.

This is not to say, however, that science is always equated with freedom and equality, as we shall see in sections 7.3.3 and 7.3.4. While attention is often given to the inherent social injustices associated with non-rational modes of thought such as romanticism, with its emphasis on the differences between people (see section 3.1.2), 'rational' science has also been known to legitimise and even exaggerate supposed differences, and to help to perpetuate inequality.

2.5 THE CLASSICAL LEGACY IN TECHNOCENTRIC SCIENCE

We now attempt to summarise the nature of modern science as it is widely perceived. This will be done by describing one scientific philosophy which considerably influences the popular view of the nature and importance of science, and by referring to some writings which, in our view, convey a picture of science that is widely shared. The legacy of classical science will be very evident, as will be the extent to which these views inform a major part of official ideology concerning social and environmental relationships. This ideology is also highly 'technocentric'. We want however to emphasise that we shall be characterising one particular view of science, and that although this view *is* very widely shared it is not the only view to be held in the past 150 years or so. There are many who have repudiated it, though not all of them have been 'anti-science'. They simply have a different conception of what science is, and of its importance - we take up the views of one such group, 'ecocentric' scientists who embrace the 'organic systems' philosophy, in Chapter 4.

2.5.1 What Science and Scientists are Like

The popular, official and technocentric conception of the nature of science and scientists is enshrined in the scientific philosophy known as *positivism*. This is the 'Philosophical system of Auguste Comte (1798-1857), recognising only positive facts and observable phenomena, and rejecting metaphysics and theism' (OED). Comte, in the 1830s and '40s promulgated the supremacy of science in seeking natural laws, and defined science as the study of 'real', that is empirically-observable (or what can be deduced from observable), phenomena. These phenomena and their interrelations were established through common and repeatable observational methods. All branches of science used standard methodology, involving a step-by-step process of accumulating knowledge and incorporating it with theory. The object of science was to make law-like generalisations which would be true throughout time and space about these phenomena and interrelations. Science, then, was concerned with the *nomothetic* (what is common and governed by laws) rather than *idiographic* (individual or unique, and therefore non-repeatable) features of things. The point of establishing laws was that they could be used to *predict* what would happen, given particular circumstances, and prediction led to the ability to *control*. Thus scientists could modify nature for 'society' through manipulating the variables in *deterministic* relationships (cf the Baconian creed). Furthermore, Comte believed that human society was amenable to prediction and manipulation via social science. This view followed Locke's advocacy of the application of science to society, and it has prevailed in modern times, as witness the proliferation of the social sciences in Western culture. The assumptions of positivist social science, characterised by Johnston (1983 p26), seem to be firmly based upon those principles which were enunciated by the chief figures in the scientific revolution. They are: social events involving human decisions have identifiable and verifiable causes; decision-making accords to a set of laws to which individuals conform; there is a world of behaviour which can be objectively seen and recorded; social scientists are distinterested professional observers, detached from subjects they might be involved in in other areas of their lives; as with inanimate matter, society is structured, and the structure changes in response to observable laws (this is equivalent to functionalism, see 1.5.2).

We may note that this unemotional, detached scientific view of society - which Johnston denies is crudely mechanistic - is very much a technocentric appraisal. It is applied whenever the technocentric has to make decisions about the man-made (i.e. urban) environment. Its reductionist overtones are strongly inherent in such planning techniques as cost-benefit analysis (see Sandbach 1980). To those who would argue that decision-making involves the uniqueness of human beings, and that deterministic laws cannot be applied to creatures with free will Johnston replies: (a) that the uniqueness represents a unique combination of circumstances governing the operation of relevant laws rather than singular situations incomparable with others: and (b) that 'all behaviour and beliefs are ultimately determined' - the problem is the inability of social science to identify the causal chains, but a 'random error' element can be introduced into the calculations to accommodate what is now interpreted as 'free will' (cf. Moss 1979 - see 4.3.2).

This statement reflects a widely-acknowledged limitation of modern science, which is that prediction cannot be based for all sciences on the same degree of mathematical certainty which might obtain in, say, classical physics. Therefore the aim, in natural sciences like biology, geomorphology and geology, governed by multivariate relationships, is to establish the levels of *probability* of given outcomes with increasing accuracy as a basis for prediction. (Here, though, the very modern, as opposed to classical, physics seems to take a contrary position to earlier certainties, being based very much on probabilities.)

In questions of the relationship between science and society it is not only the problem of certainty, but also that of 'objectivity' which particularly arises. The idea that the scientist must be a detached observer - which hinges on the subject-object/primary-secondary-qualities distinction elaborated by Galileo and Descartes - is very apparent in the assumptions of positivist social science listed above, and it pervades natural sciences too. There are obvious shortcomings in it, of which the positivists themselves were aware:

> Unlike religion it [science] is ethically neutral: it assures men they can perform wonders but does not tell them what wonders to perform ... The philosophies that have been inspired by scientific *technique* are power philosophies, and tend to regard everything non-human as raw material. Ends are no longer considered: only the skilfulness of the process is valued. This is also [like unmoderated romanticism] a form of madness (Russell 1946).

Schumacher (1973) summarised this feature as the tendency for modern science (and scientific education) to be immersed in questions of how to achieve technical feats (*know-how*) while neglecting questions of what feats it would be morally right and wrong to achieve (*know-what*). The separation of these two categories of problem from each other is characterised by Skolimowski (1981) as the divorce of *knowledge* from *values*. In his view it is a separation which took place in recent history, with the scientific and industrial revolutions. Our examination of medieval cosmologies (2.2 and 3.1) supports this view, for it shows that in them knowledge and values were intimately linked (if anything scientific knowledge had to be subordinated to religious values).

A clear example of the ascendancy of 'know-how' over 'know-what' is given by the case of the atomic scientists of the 20th century, of whom Robert Jungk asks 'Why are we interested only in what scientists do and not in what they are?' Jungk maintains that 'Nothing less than the arbitrary and unnatural separation of scientific research from the reality of the individual personality ...' (that is, the subject-object dualism) 'could have allowed the creation of such monstrosities as the atomic bomb and the hydrogen bomb'. He describes an encounter with a brilliant mathematician in a Los Angeles street:

> His face was wreathed in a smile of almost angelic beauty. He looked as though his inner gaze were fixed upon a world of harmonies. But in fact, as he told me later, he was thinking about a mathematical problem, whose solution was essential to the construction of a new type of H-bomb ... this man had never watched the trial explosion of any of the bombs which he had helped to devise. He had never visited Hiroshima and Nagasaki, even though he had been invited ... To him research for nuclear weapons was just pure higher mathematics untrammelled by blood, poison and destruction. All that, he said, was none of his business (Jungk 1982 pp13-14).

Since, and substantially as a result of, the development of these weapons, there has been some move towards 'social responsibility', or subjectivity, in science even though, as Jungk describes, too little was done too late to prevent major escalations of the arms race.

We have suggested that generally there is a lack of social responsibility or awareness among scientists, and that this is partly a legacy of the classical idea of detached objectivity. But it is also true, as we have shown, that Bacon's man of science was not divorced from the social implications of his discoveries (in *New Atlantis* he decided what new knowledge should and should not be passed on to the people). The more modern trend towards 'narrowness' and an inability among scientists to place their work in a broader social context is blamed by Ravetz (1971) on the tendency for scientific research to become de-skilled through the division of labour and the application of a managerial approach (see also Hales 1982 and Albury and Schwartz 1982). He bemoans this tendency, and revives the Baconian view that the scientist must be an accomplished craftsman.

He also points out that the general public do not see the deficiencies of modern science and scientists. To them 'In a variety of ways the pursuit of science ... has functioned as a religion', as Bacon foretold. The public conceives of the scientist in pursuit of a truth which will live on after death, and these truths are, in keeping with the positivist philosophy, regarded 'to an astonishing degree as hard and reliable'. Richards (1983 p148) endorses Ravetz's view, maintaining that many practising scientists cling to an idealised conception of their profession, propagating a view of scientific 'truth' which implies complete certainity, objectivity and detachment, though they know that most modern science is paid for by governments and businesses and is directed towards specific economic and political interests. 'Incompatible as these two positions may seem, it yet remains true that the prestige and authority of science is such that they are widely accepted, more or less unthinkingly, by the public at large. To some degree, at least, scientism has usurped the territory formerly held by religion'.

And *The Economist* (1981) in an article on the nature of 'knowledge' spreads unthinkingly the new faith

> What should be beyond argument is that there is an accretion of known facts. On the whole science is 'true'. To deny that man knows more about the workings of nature now than he did in the Middle Ages is perverse ... the alternative to accepting that there is a strong measure of truth in science is to go back to blaming a witch doctor when a cow is sick.

Following Bacon again, this article takes a cumulative view of knowledge: 'today's scientists stand on the shoulders of their predecessors to place new bricks on the pyramid of knowledge ... Take Copernicus out of the pyramid and you have to rethink Einstein'. Furthermore, empiricism is the means of making man (and science) less fallible. For man's mental facilities are like those of animals; though more developed they are, too, inaccurate, but this can be minimised by seeking accurate measurement, experiment and the use of theory. And the relative truth or falsity of theory and fact in science should be weighed by 'the collective scrutiny of other scientists'. Merton (1972) has also echoed Bacon, in defining the scientific community's ethos as directed towards a universal good through disinterested pursuit of a 'truth' that is independent of society at large. Richards, however, points out the myth of this perspective, popular though it might remain. For

20th-century science is typically 'applied', and practised in the government or industrial laboratory by big teams of increasingly specialised personnel. These communities, quite unlike the Baconian model, keep their knowledge secret, and are loyal to the ideals of financial soundness and a non-egalitarian, hierarchical organisation. Far from having a sense of wider social responsibility, many of these scientists, according to a 1969 survey, are 'unaware that pure science might have a moral ethos of any kind' (Richards p122).

2.5.2 The Importance of Science

One widely prevalent view of science - that is the kind of science which has developed as the child of the scientific revolution - is that it is *all-important*. Scientific method, scientific rationality, and the products of applying science are regarded as providing the means to, and indicators of, social progress. What is 'scientific', and what is produced through science has therefore come, in this view, to have a far greater validity than what is non-scientific. The view is enshrined in the development of Comte's positivism which is known as *logical positivism*. It was formulated in the 1920s and '30s by a group of philosophers (of which Russell was a member) known as the 'Vienna Circle'. They expounded logical empiricism, where rules of logic were used to analyse, or break down, propositions about the world into elementary statements, or *hypotheses*, which could be empirically tested against reality. But they were concerned with doctrine as well as method. Their doctrine held that knowledge which was not empirically verifiable (such as intuitive or emotionally- or spiritually-derived knowledge) was not as valid or meaningful as that which was empirically verifiable (cf. Descartes). It held that the use of reason, which was amenable to scientific method, should form the basis for social action (cf. Bacon), and that judgements based on such scientific reasoning were objective and superior to subjective judgements. Values, emotions, intuition - all those facets of human life which could not be empirically *verified* (or falsified) i.e. 'proved' through observation and measurement and logically argued - these should be excluded from valid knowledge. That is, they should not form the basis for action. Richards sums up the tenets of this philosophy of 'scientism'. It sees science as 'truth institutionalised'; thus to append the word 'scientific' to one's evidence is 'to lend one's argument weight of a special kind' - it is 'to encourage the view that it is altogether respectable and must be taken seriously'. And the defining characteristic of scientific laws is that they describe 'relations and regularities that are invariable', while scientific knowledge is universal in time and space. Empirical science is therefore uninfluenced by cultural factors, such as dogma or mere opinion - it emphasises what can be agreed by all observers, having been derived from precise measurements performed by impersonal instruments.

From this philosophy there stems, nowadays, the kind of elitism which seems to hark back to Bacon's conclusion that decision making should be the province of a scientific priesthood. For if scientific knowledge is so superior to alternative kinds of knowledge, then it follows that those who are most versed in the superior knowledge should have most say in decisions (e.g. about environmental issues).

Ravetz conveys the view accurately. He first details how public appreciation of science is needed to further research - which is a 'continuous disciplined advance from the known into the unknown' (cf. the 'pyramid of knowledge', and 'truth as the daughter of time'). But this appreciation is based on the ability of science to produce labour-saving devices and strange and wonderful effects, as in space travel. The latter (dubbed the 'Gee-whiz' factor by Hales) is a sort of revived belief in 'natural magic'. But what the public does not appreciate - and is *incapable* of appreciating - are the natural laws which produce the 'magic'. This being so 'The layman must trust the scientist as a man whose work, while incomprehensible, is genuinely worthwhile for its own sake'. This is *exactly* the type of statement which today often emanates from 'experts' in the nuclear industry, or from military- or economic-science experts, as they each attempt to persuade the general public of its unfitness to participate in decisions about nuclear power, nuclear weapons or the way to reduce unemployment. Its logical conclusion is summarised in Ravetz's observation that there are 5-10% of the public who, while not doing research, have nonetheless studied science as students, and that:

> It is widely hoped that a higher proportion of places of responsibility in public life will be taken by such people so that there will be an intelligent and competent mediation between the scientific community and the general public.

Kahn and Weiner (1968) also appear to point to the scientist as the person most fitted to control society. For it is the scientifically-based elite which most respects the 'disastrous potential' of the 'Promethean accomplishments' of modern science and technology. This power to subdue nature to human will is the result of 'eons of striving', which is on the verge of success, for 'as we approach the beginning of the 21st century our capacities for and commitment to economic development and technological control over external and internal environment ... seem to be increasing and without foreseeable limit'. Clearly they feel that this Baconian ideal is worthwhile, for 'To increase economic development is to increase the availability of at least some of the things that people want and need'. On the whole technological and scientific developments are 'occasions for satisfaction', and Kahn and Weiner share the widespread belief that the power to change the environment constitutes 'progress', and it is not feasible or desirable to halt it. But with the increasing scale of mastery over nature, the penalties for making mistakes become larger. The only way to avoid them is to moderate 'Faustian impulses to control the environment' via the centralisation of political, economic and technical power. And the only way to do *this* is to '...somehow arrange matters so that the inescapable increase in regulation of human choices remains in the hands of people who will respect its disastrous potential'. These must be primarily the scientists; they will constitute the protectors of the 'freedom of dangerous choices', either directly or through their political influences.

Herman Kahn was one of the most optimistic of technocentric optimists. He argued vigorously against those who would limit economic growth, saying that only unbridled growth would bring social progress. Central to such growth, and therefore synonymous with social progress, was technological progress which would be informed by scientific advances. Kahn's 'scenario' for the period 1980-2000 was an increasingly prosperous one:

> The growth of world markets for food, the ability to introduce technology effectively into lower-income countries, and the availability of cheap labor, combine to create an environment in which growth spreads throughout the world. One nation after another joins the club of newly industrialised societies (Kahn 1980).

The prosperity would hinge on technological developments such as a 'world-wide capacity for modern industry and technology', 'worldwide green revolution', an enormous expansion (to the point of surplus) of mineral and energy reserves through technical progress in exploration and exploitation, and selected improvements in environmental quality. Kahn was euphoric about the benefits these would bring, povided that 'the forces against progress' (those who wanted planned growth or no growth) continued to be held in check:

> Although many of the difficulties facing the world are serious, many of them are but growing pains in a world half-way through the Great Transition. The Great Transition will last 400 years and will ultimately yield enormous benefits from the spread of progress and greater ability to control technology for the good of man. Along the way there will be both reactions to progress and new adventures such as those beneath the oceans and in space. At the end of the journey there will be prosperity and self-fulfilment on a scale the world has never known.

Such a cornucopian statement would have been worthy of Bacon, Condorcet or Locke - all firm believers in progress through the application of rationality to the task of manipulating nature. And, as Ravetz says, this belief persists: 'The success and vigour of scientific research, and the effectiveness of its technological application are generally accepted as indicators of the quality of a nation's life' (cf. Bacon). As for the general public's appreciation of science and the scientist, this:

> leaves nothing to be desired. We are all grateful for the comfort and security of life that is achieved by modern technology, and prepared to accept the claim that all these good things are a product of scientific research. Moreover, in the common-sense understanding of man's relation with his environment ... 'Science' reigns supreme. [New scientific technology has the power] for transforming the lives of men, women and children, now helping them to emerge from what Snow calls 'the great anonymous sludge of history' (Ravetz 1971).

This should be compared with Timothy Walker's (1831) *Defence of*

Mechanical Philosophy, written to justify science in the face of Thomas Carlyle's attack on it in *Signs of the Times* (1829). Separated by 140 years, nonetheless both Walker and Ravetz and Kahn encapsulate very much the technocentric assumption that science equals progress and that this is not to be challenged by anyone of right mind:

> Fitted up, as it [the earth] now is, with all the splendid furniture of civilisation, it no more resembles the bleak, naked, incommodious earth, upon which our race commenced their improvements, than the magnificent palace resembles the low, mud-walled cottage. From the effect, turn your attention to the cause. Examine the endless varieties of machinery which man has created. Mark how all the complicated movements cooperate, in beautiful concert, to produce the desired result ... We do entertain an unfaltering belief in the permanent and continued improvement of the human race, and we consider no small portion of it, whether in relation to the body or the mind, as the result of mechanical invention.

And, says the technocentric, if this progress has now brought undesirable environmental side-effects, these should be eliminated not by abandoning science, but by using it to a greater extent. The problems should be reduced to problems of economic science, and they must be approached with optimism:

> What is needed now is a fresh approach which sees environmental problems basically as economic problems which tell us not that we are living at the end of the world but only that we are doing a few things wrong which need to be corrected. Environmental problems must be approached in the spirit that they are solvable, and not that they are messages which convey the malignant intent of an evil business establishment, or the first reckoning of domesday. A new pragmatic, optimistic approach to environmental problems would be the best possible legacy of the Age of Environmentalism (Tucker 1981).

2.5.3 Is the Popular Image of Science Breaking Down?
This technocentric view of nature and society, based very much on classical science, is the one which, as we have said, has permeated Western popular consciousness - our cultural filter - during the past two to three hundred years. Much of the rest of this book will show, directly or indirectly, how it is open to criticism. There are those who criticise scientific rationality along romantic, anti-rationalist, anti-mechanistic lines (Chapter 3). They reject the metaphor of nature as a machine, they reject the Cartesian dualism and the positivist relegation of 'secondary qualities' in humanity and nature. They see 'science' as hard-headed and cold-blooded, and advocate a highly emotionally-charged 'counter culture' (Roszak 1970). They reject reductionism, too, and want to see man and nature in a holistic, dialectical relationship involving the material and spiritual. These critics are not only 'ecocentrics'; they are also drawn from those who speak from a more clearly-defined materialist and political (left-wing) perspective (Chapters 5 and 6). Their attack on classical science is an account of what it does to encourage an atomistic and de-humanised society. Science, as they see it, always has been class-allied (Gorz 1976) and has developed in

response to the specific needs of capitalism, which do not accord with the needs of most people, viewing them as *things* - units of consumption and production. As such, science cannot be trusted to give us an impartial view about our relationships with each other or with nature.

Both of these lines of criticism have a lengthy history, as we shall show. And they have been boosted in the past two decades by a wave of more popular mistrust of science, as expressed, for example, in the environmentalist movement which we have reviewed in Chapter 1. This wave has represented a backlash against the technological optimism - the 'white heat of the technological revolution' - of the 1960s and the early 1980s. It has not been founded, however, merely on irrational feelings. There are many legitimate 'hard' questions about whether technocentric science can 'deliver the goods' on its own materialistic terms. Pollution and resource depletion, the threats of nuclear war or massive unemployment - all these dangers have not been seen to have been abated through the application of more science and management - as they should be according to technocentric theory. Indeed, they are often perceived to stem *from* science and technology. Thus the equation between science and progress in the popular view may be starting to break down. And the image of a competent and altruistic priesthood which knows best has become considerably tarnished. A study of the nuclear industry's development over the past 30 years would show this *par excellence*. For example, the Baconian 'family' of scientists who developed in the 1920s and '30s the theoretical physics which were needed to unleash atomic power soon broke up after the Bomb was developed. Heisenberg remarked that in 1939 'twelve people might still have been able, by mutual agreement, to prevent the construction of atom bombs', but they let the opportunity go. Another physicist, Weizsacher, commented: 'The fact that we physicists formed one family was not enough. Perhaps we ought to have been an international order with disciplinary power over its members. But is such a thing really at all practical in view of the nature of modern science?' (Jungk 1982 pp81-2). Today, buoyant optimism seems to have been replaced more generally by this defensive pessimism, and notions of benefit for all have been rudely upset by an obvious power struggle over the issue of future energy supply involving sectional interests in and outside the nuclear power industry. Technocentrics, in all this, have frequently been seen to get their assumptions and their calculations wrong - in 'scientific' terms.

> In analysing complex, real world processes ... scientific analysis has proven to be extremely limited. One need look no further than a review of past predictions of nuclear electricity costs and nuclear power growth to see how badly scientific method can fail: since the mid-sixties, estimates of the cost of nuclear energy have risen sevenfold, more than tripling after adjustment for general inflation ... Don't these enormous errors by supposedly competent analysts create in you any tinge of doubt about the efficacy of scientific analysis in public policy formulation? (Taylor 1980, writing to a firm believer in the superiority of 'objective', 'scientific' analysis for solving the world's problems.)

CHAPTER 3

THE NON-SCIENTIFIC ROOTS OF ECOLOGICAL ENVIRONMENTALISM

3.1 PRELUDE: PLENITUDE AND THE GREAT CHAIN OF BEING

3.1.1 Pervasive Ideas

The main tasks of Chapters 3 and 4 will be to demonstrate that modern ecological environmentalism (ecocentrism) has particular affinities with 19th-century romanticism - characterised as a 'non-scientific' root - and also the ideas of Malthus and Darwin - characterised as 'scientific' roots. However, before proceeding to these tasks it is as well to consider two powerful and related pre-19th-century ideas on the relationships between living and non-living objects on earth, and between earthly and 'heavenly' beings. These ideas of the *Great Chain of Being* and *plenitude*, superficially resemble some of the ideas of romanticism and modern biology, and thereby are linked with ecocentrism. We shall describe the ideas, consider their apparent links with ecocentrism and, finally, question how strong such links really are.

The ideas originate with the Greeks, and they constitute a neoplatonic cosmology, which was expounded principally by Plotinus (204-270AD) and summed up by Macrobius in the early 5th century. He gave it the form in which it was transmitted to medieval writers and became adapted to and part of their cosmology - the biological counterpart of the physical scheme discussed in Chapter 2. According to Lovejoy (1974), despite waning faith in metaphysics and increasing faith in Baconian empiricism, the Chain of Being cosmology was widely diffused and accepted as late as the 18th century. John Locke, for example, restated it in his (1690) *Essay Concerning Human Understanding*, and late-17th-century scientists spent much time discussing which monarchical animals - lion, eagle, or whale - were at the top of the hierarchies into which they were assumed to be organised (Thomas 1983).

We may again note here that, as with the cosmology described in Chapter 2, the paradigm for medieval science was essentially set by theology. As Lovejoy describes it, for most men of science 'the theorems implicit in the conception of the Chain of Being continued to constitute essential presuppositions in the framing of scientific hypotheses'. And this continued to be so into the 18th century, while Thomas reports that as late as 1834 a zoologist named William Swainson thought that zoology's aim was to discover animals' stations in the scale of creation. In these ideas, then, physico-theology remained very much alive, despite rationalism, the successes of Baconian science, and opposition from Enlightenment *philosophes*.

Lovejoy (p59) says that the ideas constituted:

> A plan and structure of the world which ... many philosophers, most men of science and, indeed, most educated men were to accept without question - the conception of the universe as a 'Great Chain of Being', composed of an immense, or ... of an indefinite, number of links ranging in hierarchical order from the meagrest kinds of existents, which barely escape non-existence, through 'every possible' grade up to the ... highest possible kind of creature, between which and the Absolute Being the disparity was assumed to be infinite - every one of them differing from that immediately above and that immediately below it by the least possible degree of difference.

Macrobius' summary emphasises the importance of interconnections:

> Since, from the Supreme God Mind arises, and from Mind, soul, and since this in turn creates all subsequent things and fills them all with life, and since this single radiance illumines all and is reflected in each, as a single face might be reflected in many mirrors placed in a series; and since all things follow in continuous succession, degenerating in sequence to the very bottom of the series, the attentive observer will discover a connection of parts, from the Supreme Good down to the last dregs of things, mutually linked together and without a break. And this is Homer's golden chain, which God, he says, bade hang down from heaven to earth (quoted, Lovejoy p63).

Alexander Pope stressed interdependence:

> From Nature's chain whatever link you strike,
> Tenth, or ten thousandth, breaks the chain alike.

Thus, to eliminate one link - one creature or one part of inanimate matter - would fully dissolve the 'cosmical order', and, ceasing to be full, the world would be incoherent. Hence living things are interlinked, via regularly-graded affinities, in an hierarchical order (Figure 5). The transitions - from species to species, from inanimate matter to animals, or from animals to plants - are almost imperceptible and quasi-continuous.

These ideas of continuity and gradation are fused with that of plenitude, or fullness. This holds that the universe is filled by diverse living things, such that no genuine potentiality of living being is left unrealised, i.e. no species that *could* theoretically exist does not in fact exist. Not only is there diversity, but there is also an abundance of creatures, stemming from their theoretically-infinite fecundity (which was curbed only by competition and nature's limits, see Section 4.1, on Malthus). Plenitude, too, held that the world was better the more it contained.

69

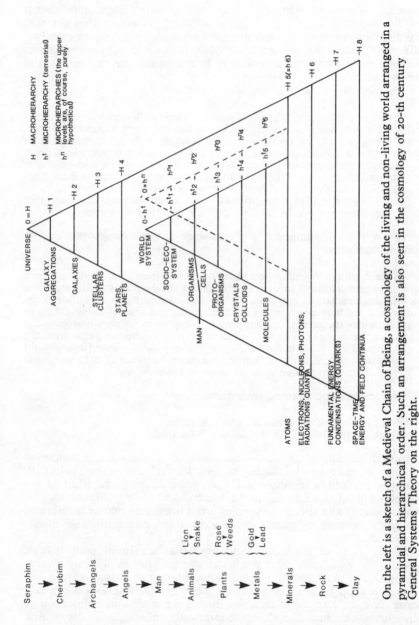

FIGURE 5   Chains of Being (source for systems diagram: Laszlo 1983)

The link between plenitude and the Chain of Being was that at the top of the hierarchy was a transcendent *hypostatis* (personality or underlying substance) - a First Principle: The *One* or *Good* in the Greek version, which became *God* in the Christianised version. Emanating from this was a Universal Soul which ruled the universe at two levels; a higher spiritual and a lower material one. Being perfect, the transcendent First Principle would not remain in itself but *spilled over*, generating in plenty the things below it in the chain - the soul, the archangels and angels, man, animals, plants, metals, stones, mud and slime. This hierarchy of being was all, then, an emanation from the overflowing fullness of the One, or God, This was why all things were linked and interdependent as an *organic whole*.

3.1.2 The Links with Romanticism and Ecology
Romanticism
There are two ways in which plenitude and the Chain of Being appear to link with romanticism. One is advanced by Lovejoy through a seemingly tenuous argument. He maintains that, like romanticism, the Chain of Being idea ran counter to rationalism and the search for *universal* laws and standards to govern and explain society and nature (see Chapter 2). Whereas Enlightenment reformers aimed at universally-shared social standards, coupled with the democratisation of material wealth through industry and technology, the Chain of Being idea encouraged the belief (from plenitude) that *differentness* and diversity equalled excellence. Similarly, romanticism gloried in uniqueness and diversity; in many literary genres and verse forms, in the demand for local colour and particularised and faithful landscape descriptions, in distrusting universal political formulae and in championing the individual.

Consequently Lovejoy sees Plotinus as the key to romanticism, and echoed in Schiller, who enunciated plenitude and said that the Great Inventor would not allow even error to remain unused. So *all possible modes* of being and activity, perfect and imperfect, constitute the romantic goal. Romantics seek material and sensuous experience. Schiller saw this as the 'sensual impulse' which, as Lovejoy points out, can lead either to catholicity and tolerance of others, or to its opposite in the pursuit of personal, national and racial idiosyncracies (see 7.3.1).

The second link with romanticism is more obvious. It is with the 'bioethic' - the respect for all life forms in their own right (rather than just for pragmatic reasons). This was enunciated particularly by the American romantic transcendentalists (see 3.2), from Thoreau through to Leopold (1949). The Chain of Being gave man several good reasons to feel humility in respect of the rest of nature. First, the removal of *any* link from it would destroy the chain, hence all links were equally important for the Chain's completeness. Secondly, man was situated only in the middle of the Chain, at the point of transition between being purely animal-like creatures of instinct and physical senses, and thinking beings with a soul who transcended the physical. Superior to him were the intelligences of the angels and, according to 17th- and

18th-century writers, the beings of Jupiter, Saturn, and other worlds which were further away from the sun than earth - their superiority derived from their lesser dependence on this inert body (note the correspondence with the cosmography described in 2.2.1). Third, man's distinctiveness from the creatures below him was hardly chasm-like because of the continuity principle which led to almost imperceptible transitions.

Thus Lovejoy (p200) concludes that despite other strains in the 18th century which worked against humility, 'and prepared the way for those disastrous illusions of man about himself which were to be so characteristic of the century that followed ... the immense influence of the complex of ideas which was summed up in the cosmological conception of the Great Chain of Being tended ... to make man not unbecomingly sensible of his littleness in the scheme of things'. Similarly Nash (1974) thinks that the Chain made synthetic and absurd the idea that nature was subservient to man, but also that few had drawn these conclusions before the 19th century.

Related to the element of humility towards and respect for nature, which is common to both Chain of Being and bioethic, is that of *animism*. Animism is the attribution of living souls to animals, plants and inanimate objects. According to Mills (1982), the organic analogy which is inherent in the Chain of Being idea (see 2.1, and below under 'Ecology') is essentially animistic. If the cosmos is an 'organism', stemming from the immanence of the One - the Absolute Being which is in everything - then it is but a small step to have nature and natural objects endowed with the attributes of organisms, especially humans. This was done not only by giving, for example, 'brows, shoulders and feet' to mountains, or 'heads, gorges and mouths' to rivers, or a circulatory system to the whole earth, it was also possible to give the parts of nature some of the universal spirit, which was held to reside in them (this was *pantheism*, in which 'God is everything, and everything God', OED).

Animistic and pantheistic ideas appeared not only in classical to Renaissance writing, when the Chain of Being concept allied to them was pervasive; they also seem to have been carried forward into the romantics. Thomas (1983) documents an obsession with trees from 1770 to 1850 which made it impossible to destroy them without strong regret, based partly on the idea that trees feel pain like humans. He traces this sentiment through to many poets, including Wordsworth and Tennyson, and calls it 'partly religious'. Romantic poets preserved the classical conception of woods as the haunts of sylvan deities, and the notion of them as a place for solitude and meditation seems to have been reinforced by the idea of nature as a religious force to be worshipped (see section 3.3.1). Romantics, especially the American transcendentalists, saw trees as God's temples. In so doing they appear to have carried on from the 17th-century practices of worshipping plants, talking to them and offering them libations, which are documented by Thomas. These in turn strongly resemble the 'pagan animism' described by White (1967), in which every tree, spring, hill, and stream had its own guardian spirit which was accessible to but unlike them - centaurs, fauns, mermaids. 'Before one cut a tree, mined a mountain, or dammed a brook', says White, 'it was important to placate the spirit in charge of that particular situation...'

So there are parallels between 'pagan', classical, Renaissance and romantic animism, which suggest that the idea, like pantheism, continually resurfaces over time. It resurfaces, too, in ecocentrism, wherein for instance ecological writers (e.g. McHarg 1969) point to the animism of 'primitive' peoples such as the Navajo Indians as examples for 'advanced' cultures to emulate. The theme is strong in Callenbach's *Ecotopia* where trees are given a heart-to-heart talk before felling to explain and apologise for the regrettable circumstances in which the atrocity is to be perpetrated.

Ecology

Lovejoy maintains that the Chain of Being provided a paradigm for much 18th-century biology. Whether this claim can be extended to embrace 19th and 20th-century biology is, as will be seen in the next section, debatable. Nonetheless, there are thought-provoking resemblances, (see Figure 5) if not historic, links between facets of the Chain of Being idea and many modern ecological concepts, such as interdependence and systems, evolution, niche and diversity, hierarchy, and the determinism which stems from them.

Thus, the acknowledged importance of each link in the Chain as vital in maintaining the whole structure suggests the notion of *interdependence*, which is a crucial feature of all parts of nature in an ecological view. The Chain's links are provided via the Universal Soul which emanates from the Supreme Being. The Soul is the immanent force linking the parts of the whole, and the world is in a way a *single life* stretched out to an immense span and consisting of *linked parts*. One is reminded here of how some modern systems advocates also present the idea of an immanent ubiquitous unifying force in the universe, which is *energy*. All systems are united by the flow of energy between them, and the material comprising them is, in a sense, 'frozen' energy, which is being constantly transformed - from matter to energy and back again.

Both animate and inanimate nature were included in the organic conception of the Chain. Thus an organic analogy was made, whereby stones were thought to be capable of growth and reproduction, and therefore the exploitation of the rocks via mining - hewing stones from the 'living rock' in the 'womb' of 'mother earth' - was regarded as tantamount to abortion. Minerals were thought to mature, into gold, and the earth had a circulatory system equivalent to:

> the blood in animals which is always moving from the sea of the heart and flows to the top of their heads; ... as one may see when a vein bursts in the nose, that all the blood from below rises to the level of the burst vein. When the water rushes out of a burst vein in the earth it obeys the nature of other things heavier than air, whence it always seeks the lowest places (Leonardo da Vinci, quoted in Mills 1982 pp244-5).

Here are similarities with Capra's organic systems view (see section 4.3.4) in which the whole earth is seen as a living organism, making the systems links more intimate than if they were merely mechanical. Capra says (1975) that the compatible eastern concept of the unity of all things is personified by Shiva, the Cosmic Dancer, 'the god of creation and destruction who sustains through his dance the endless rhythm of the universe'. Shiva is perhaps analogous with the One; the Good at the head of the Chain.

Evolution, says Lovejoy, is suggested by the notion of continuity in the Chain and the gradualness of transition between one link and the next. Since, however, the Chain was for the most part a static concept, while evolution is a dynamic one, this would have applied, if at all, particularly after the Chain was 'temporalised' in the 18th century, that is, held no longer to be fixed in time and a result of the simultaneous creation of all the links, but to be changing.

Resemblances also suggest themselves between the plenitude idea and the ecological concepts of *niche* and *stability-in-diversity*. Niche conveys the notion that each place on earth that has appropriate conditions will have acquired some species which will have developed to fill it. And the modern dictum that diverse communities are more stable than those comprising only a few species parallels the old one which proclaims that the world is better the more it contains.

The structure of the Chain is clearly hierarchical (Figure 5), as are many biological and botanical classifications. Indeed, the concept of *hierarchy* is strongly conveyed in modern ecology, via, for example, the ideas of food chains and predator-prey relationships. Such relationships are held to be natural and inevitable (and by implication acceptable) in the natural world. They are deterministic (see 4.3.3), for it is not within the power of any species to alter its preordained position in the food chain.

Lovejoy draws our attention to the determinism of the Chain of Being. It supported the doctrine of *theodicity*, which vindicated divine providence, i.e. 'excused' God for permitting evil to exist. For if plenitude holds that the best of all possible worlds is a *full* one, then the world must contain evil as well as good. From there one could argue that the inferior station in the Chain, and the concomitant suffering, of non-rational sinless animals were necessary for the good of the whole assemblage - it was in the nature of some to be eaten; they were needed to make up the set. Lovejoy quotes Plotinus, who said 'We ought not to demand that all should be good, nor hastily complain because this is not possible', Augustine, who said 'If all things were equal, all things would not be', and Pope, who said in his *Essay on Man*:

> Order is Heav'n's first law: and this confest,
> Some are, and must be, greater than the rest,
> More rich, more wise.

Social implications are obviously drawn here - they are that any demand for equality runs contrary to nature. *The unequal order is pre-determined*, so there is nothing one can do about it, or ought to try to do.

These ideas shade into the common 18th-century belief in the inherent superiority of some races and peoples over others. They sometimes relied on principles of continuity between species, which meant that the differences between man and the apes must be finely graded. Therefore, as Lovejoy describes, 18th-century biologists spent much time in searching for 'missing links'. Just as the Hydra was seen in 1739 as the link between animals and vegetables, so the Hottentots became one of those unfortunate peoples who were branded as 'savages', little better than apes and the link between apes and 'man'. In fact the gap between Hottentot and ape was regarded as smaller than that between the former and the white Anglo-European intelligentsia.

3.1.3 Links, or a Discontinuity?

As indicated, it is tempting to draw continuous threads running from the pre-19th-century concepts of plenitude and the Chain of Being, through the 19th century and into 20th-century ecocentrism. The temptation to do the same with the implications which were drawn for man from the Chain of Being is also strong - as we shall see in Chapter 7, there is a strain of ecocentrism which draws upon ideas of social hierarchy and determinism, and postulates the existence of inherently superior and inferior elements in society.

But it is precisely because of the *socially* reactionary nature of its concepts that the Chain of Being idea was, according to Bynum (1975), abruptly discarded. Its deterministic, hierarchical and racist ideas, he says, appealed to the *status quo* in society, but the 18th and 19th centuries were characterised by a reforming zeal. Because reformers found them antagonistic to their aims, the ideas were despatched to the garbage can of history. Methodists and other evangelicals, slave-trade abolitionists, Bentham and the Utilitarians - all were part of the generations which discarded the Chain of Being. They were, says Bynum, conscious of achievement, exhilarated by change, and fired by the idea 'that man could harness nature'. Modern biology also had no more use for the Chain, for such as Lamarck and Cuvier had demonstrated that between the animate and the inaminate, plants and vegetables, brutes and men there were clear discontinuities and no gradations.

If this view is correct, then Baconian scientific rationalism had finally won a victory over older medieval theological science. The challenge to the philosophies of classical science then came not from the theologists with their Chain of Being but from the 19th-century romantics. However, with the latter's strong propensities for animism, a view of nature as organic, and a humility towards it which reflects man's lowly station in the cosmic order, it seems likely that at least they, if not the scientists, may have to a degree carried foward into 20th-century ecocentrism some of the concepts inherent in the Chain of Being cosmology.

3.2 THE BASIS OF ROMANTICISM

(The remainder of this chapter is based partly on material supplied and interpreted by Martyn Youngs.)

Though the term 'romanticism' has a complexity of meanings and nuances, we use it here to denote the 'content and character of the Romantic movement' of the 18th and 19th centuries (Williams 1983A). It would be impossible in this space to convey comprehensively the philosophies and flavours of this movement, but some brief discussion of them is unavoidable, since romanticism, mediated particularly by the romantic transcendentalists of mid-19th-century America, greatly nourishes the 'ecocentric root of modern environmentalism' (O'Riordan 1981A), as we shall show in 3.5.

Romanticism is sometimes described as an artistic and intellectual movement, commonly finding expression in literature, music, painting and drama. However it should not be thought of as simply a set of ideas unrelated to what was happening in the material world. For it can be seen clearly as a reaction *against* material changes in society - changes in the mode of production which can be regarded as part of the emergence and expansion of industrial capitalism in the 18th century, following on the establishment of mercantile and agricultural capitalism.

In this transition production became more centralised in the city, which no longer functioned merely to gather in and consume the produce of the countryside. The factory movement and mass production were founded on processes which unleashed and controlled violent natural forces. These processes, allied to the profit motive, 'degraded' and 'despoiled', as some saw it, the environment. Cities grew at unprecedented rates, and became centres of squalor and deprivation. They began to symbolise the failure of the Locke and Hume philosophies that a perfect society could be attained by permitting people to follow in an enlightened way their self interest. The upper echelons of society were also in flux - the new bourgeoisie were displacing the old landed aristocratic order (to which Byron and Shelley belonged). The movement of population from the land, allied to the rational search for economically efficient production methods - such as the division of labour, timekeeping and mechanisation - led to a spiritual alienation of the mass of the people from the land and from each other. They became units of production, or cogs in a productive machine - they were objectivised, they and their labour were reduced to the status of a commodity.

With their affinity for the old aristocratic order, romantics had no empathy with industrialisation. Their revolt against it was born of their own material social considerations, but since the new order was increasingly *materialistic* (in the sense of preferring material possessions and comforts to spiritual values), the revolt took the form of a rejection of materialism. As Russell (1946 p653) put it, 'The romantic movement is characterised, as a whole, by the substitution of aesthetic for utilitarian standards'. Romantics noticed and hated the way that industrialisation made previously beautiful places ugly, and they rejected the vulgarity of those who made money in trade. With the emergence of new social groups, romantics sought their own self-definition in this eschewing of the commodity society. They separated themselves from the vulgar bourgeoisie and working-class proletariat not only by physical means but also by promoting the idea that their own labour - unlike that of capitalism - was not reducible to commodity values. Their labour was intellectual; it was uniquely special. It was *artistic*, whereas the labour of skilled workers was *artisan*. It was not until this period that the two words took on distinctive and separate meanings. Artisans had plain utilitarian skills, whereas artists had aesthetic skills, and were distinguished by the sensitivity with which they applied them. Mechanical production and the mechanical materialist philosophy which it fostered were emphatically cast aside, for example by Thomas Carlyle in *Signs of the Times* (1821):

> The truth is, men have lost their belief in the Invisible, and believe and hope, and work only in the Visible ... Only the material, the immediate practical, not the divine and spiritual, is important to us. The infinite, absolute character of Virtue has passed into a finite, conditional one; it is no longer a worship of the Beautiful and Good; but a calculation of the Profitable.

The European romantics constituted a dazzling assortment of artistic talents. One strand which united them was a predilection for the *freedom of the individual*. The emphasis on the unique - the idiographic - on freedom expressed through the individual spirit, feelings and passions, ran counter to the 'mechanical' philosophy of scientific rationalism (see section 2.5) where the attempt was made to find laws governing nature and society through which their behaviour could be predicted. (In this conception men are not unique; given certain conditions they will repeatedly behave in a certain way.) Unsurprisingly, romanticism is attacked by the champions of rational philosophy:

> It is not the psychology of the romantics that is at fault: it is their standard of values. They admire strong passions, of no matter what kind, and whatever may be their social consequences. Romantic love is strong enough to win their approval ... but most of the strongest passions are destructive ... Hence the type of man encouraged by romanticism ... is violent and anti-social, an anarchic rebel or a conquering tyrant (Russell 1946 p656).

(We remark on the links between romantic ecocentrism and fascism in 7.3.2.)

Romanticism was and is the antithesis of everything 'scientific' - logical behaviour, order, authority. Indeed, it has been regarded as a sweeping revolt against rationalism and the Enlightenment. The romantics maintained that science was inadequate to explain all the phenomena with which man is confronted. They regarded these phenomena - understandable through intuitional, instinctive and emotionally-based knowledge - as the most noble aspects of being human. While scientists denigrated them, romantics elevated them. In so doing they rejected the Cartesian dualism. Subjective knowledge of, and one-ness with, nature, expressed through art: this was a superior form of knowledge to that of objective, empirical and coldly-calculating classical science. Romantics exalted in fantasy, imagination, unrepressed passion and depth of feeling. Spontaneity, inner truth and the extent to which the unique point of view of the artist was expressed - all were criteria by which they judged their work. A romantic was also, according to Edwards (1972), 'sensitive and eager for novelty and adventure', revelling in disorder and uncertainty.

In opposing science and civilisation he was afraid of the dehumanisation which they appeared to bring. This fear was symbolised by Mary Shelley's (1818) dreadful product of science, Frankenstein's monster, which ceased to be the slave of man and instead became his master and destroyer. (Russell wryly observed that even while the monster is gazing on the fruits of his murders its *sentiments* remain noble. It says 'When I run over the frightful catalogue of my sins, I cannot believe that I am the same creature whose thoughts were once filled with sublime and transcendent visions of the beauty and majesty of goodness'. The images of Hitler consumed by Wagner's music, or of Burgess' *Clockwork Orange* hero kicking old men to death to the strains of Beethoven's Ninth, come to mind.)

We should note that the dualism of the rational versus the romantic was not a new feature of Western thought. As we have seen, 18th- and 19th-century material conditions simply favoured the emergence of a 'backlash' against rationalism, but the two sides of the dualism have been there since classical times. They oppose passion and inspiration to method and discipline, content and colour to form and line, Dionysius to Apollo (the former was Bacchus, the god of intoxication, associated with instinct and passion; the latter was the sun god, who was said to be the father of Pythagoras the mathematician). This fundamental philosophical dualism is of central concern to many modern ecocentrics. Pirsig (1974) bemoans the divorcing of the romantic from the 'classical'. 'What's wrong with technology is that it's not connected in any way with matters of the spirit and of the heart ... and so it does blind, ugly things ... Lately it's become almost a national crisis - antipollution drives, anti-technological communes and styles of life, and all that'. Skolimowski (1981) similarly blames environmental and social crises on the divorcing of 'knowledge' (rational) from 'values' (romantic) (section 2.5). Even the logical philosopher Russell thought that when one side of the dualism was untempered by the influence of the other a form of madness resulted. Santmire (1973 p93) echoed this sentiment when he referred to the 'ecological schizophrenia' of the American mind. 'How can we', he asked 'so intensely adore yet so violently abuse the land of our destiny?'.

Two other tenets of romantic philosophy stemmed from its opposition to the application of rationalism to nature and man to produce the industrial, 'civilised' society as a mark of 'progress'. Since such a society was complex and sophisticated, romantics revered *simplicity* - of form (in art), of action and of ideas. Simplicity was equated with honesty, and nature was beautiful (like Schubert's chamber music) because it was simple and honest - 'Beauty is truth: truth beauty; that is all ye know on earth and all ye need to know' (Keats,

Ode on a Grecian Urn). Thus there was an interest in the folk societies of the past, which were held to have been closer to nature, simpler and more honest than modern corrupt society. This idea was summed up by Rousseau's 'Noble savage' and his dictum, from the *Discourse on Equality* (1754) that 'Man is born free but is everywhere in chains'. Inherently man was 'good', and he was born so, but civilisation corrupted and degenerated him. This concept harked back to the myth, which was revived, of the harmonious idyllic rural society of ancient Greece (Arcadia), and it amounted to the *idealisation of the past*. Such a tendency was illustrated by Keats' interest in the medieval and classical, by Walter Scott's idealisation of medieval chivalry, or by William Morris' predilection for the supposed unity of man and the land in the 'natural' organic and hierarchical society of the Middle Ages. From this belief - really a restatement of senescence (see 4.1.1) - came the back-to-nature ideal so frequently associated with the American wilderness movement.

3.3 THE ROMANTIC CONCEPTION OF NATURE
3.3.1 The European Movement
In rational thought, the Cartesian dualism meant that the kind of qualities which romantics elevated in nature - beauty, colour, majesty - were regarded as 'secondary'. They were not objectively 'real', but were the products of the human mind. Beauty was in the eye of the beholder, and it had no objective existence. But romantics denied this utterly; they held that nature had something *of its own*. Qualities like colours or beauty were not secondary but had equal status with such qualities as shape, size and motion. They, too, were primary or inherent characteristics. This is of key importance, for it meant that the romantics were ascribing a significance to nature which was not dependent upon man - *nature had an integrity of its own*. Man could relate to it in a very intimate way, but if he were to be removed from the scene nature would nonetheless go on being what it was. It is but a short step from this position to that of the modern ecocentrics which is summed up as the 'bioethic' (see 1.5.1) - the idea that nature has purpose and meaning in itself.

This move towards a respect for nature which counteracted the Baconian creed was part of a profound and broader change in the conception of wild nature which occurred during the romantic period. For up to the 17th and 18th centuries it was generally held that wild uncultivated waste was to be deplored, while the regular and symmetrical forms associated with ploughing, planting, hedging and other agricultural practices were a welcome mark of civilisation. When it became a mark of progress to tame nature, and when colour, taste and emotion become secondary qualities, then the formal and mathematical triumphed over the wild. This triumph was epitomised in the gardens of 1660s Versailles. And over a century later William Gilpin (quoted by Thomas) said that most people had found wild country in its natural state totally unpleasing: 'there are few who do not prefer the busy scenes of cultivation to the greatest of nature's rough productions'. In keeping with this sentiment, unproductive mountains had traditionally been regarded as physically unattractive - people complained of the 'desert, barren and very terrible' aspect of the Lake District, the 'hideous' Pennines and the 'hopeless sterility' of the Scottish Highlands (Thomas 1983 pp257-8).

Wales, until the late 18th century, had been similarly ignored or denigrated. Its topographic perils and climatic hazards had been exaggerated. Thus the *Gentleman's Magazine* said in 1747 that Wales was acknowledged 'a dismal region, generally ten months buried in snow and eleven in clouds'. To 18th-century rational man beauty was well-proportioned and cultivated land, and wilderness held no attraction.

But because of the very spread of agricultural cultivation, and perhaps also the improvement of communications into wild areas, the mood changed dramatically before the end of the 18th century. In the garden, geometrical precision gave way to the English style of landscape garden, which emulated the randomness of nature and was 'so informal as at times to be barely distinguishable from an uncultivated field', and 'even more remarkably, wild, barren landscape had ceased to be an object of detestation and became instead a source of spiritual renewal'. The mountains which had in the 17th century been 'hated as barren "deformities", "warts", "boils", "monstrous excrescences", "rubbish of the earth", "Nature's *pudenda*", had a century or so later become objects of the highest aesthetic admiration' (Thomas pp258-9). In hand with this were shifts in the very meaning of the words 'nature' and 'wilderness' - the former from being identified with *human* nature and reason to acquiring the ideas of goodness, innocence and that which man has *not* made, and the latter from carrying overtones of fear and hideousness to suggesting undefiled purity. The word 'sublime' began to be used to describe mountain scenery - it inspired awe and wonder by its grandeur, nobility and extraordinariness. Romantic inspiration thus came from what was grand and remote. Snowdonia became the 'British Alps' and Edmund Burke, in a *Philosophical Enquiry into the Origin of Our Ideas of the Sublime and the Beautiful* (1757), urged his readers to see the sublime in the vast, rugged, dark and gloomy; from which 'came ideas elevating, awful and of a magnificent kind'. The rugged rocks, precipices and gloomy mountain torrents had become the epitome of aesthetic experience. Zaring (1977) says that the 'rainsoaked uplands, sparsely populated and largely unploughed, were beautiful in the eyes of those who were reacting against the ideas of their father'. Merionethshire, formerly the rudest and roughest county of all Wales, had replaced the civilised county of Kent as a standard of ideal beauty. Coleridge, Shelley, Wordsworth, Southey - all went to experience the solitude of North Wales in the 1810s.

But they sought more than just earthly wellbeing. By the late 18th century, says Thomas (p260) the appreciation of nature had become a sort of religious act. Nature was seen not only as beautiful but also as morally healing and exercising a 'beneficent spiritual power over man'. Thus contact with wilderness encouraged and inspired men to *transcend* the material and reach to the spiritual plane. The mountains inspired the kind of awe and reverence traditionally reserved for God, and reminded the religious of his sublimity.

Francis de Chateaubriand wrote in 1802; 'I am nothing; I am only a solitary wanderer, and often I have heard men of science disputing on the subject of a supreme being. But I have invariably remarked that it is in the prospect of the sublime scenes of nature that the unknown being manifests himself to the human heart'. Thus, to find God and come close to him, one must find 'unspoiled' nature, his creation.

3.3.2 The American Transcendentalists
The student of North American culture must be very conscious, even in the economically-obsessed climate of the 1980s, of the importance which the average American attributes to wild nature. Each year the 'centre of gravity' of the US population shifts westwards, reflecting the operation of a crude sort of environmental determinism (Bunge 1973). If people in the overcrowded East or West Coast become affluent enough, or assured that a job awaits them they migrate to the south-west and contribute to the mushroom growth of cities like Colorado Springs, Phoenix or Albuquerque. They go principally because they appreciate the same sorts of things which the romantics of the 19th century appreciated - sunshine, pure air, sparkling rivers, grand mountains and forests, and the opportunity to be alone in the wilderness. They thus carry on the tradition of the American transcendentalists, such as Muir, Thoreau and Emerson. We should remember that, as in Europe, the love of wilderness which they expressed (partly) displaced earlier very opposite feelings towards nature. As Nash points out, wilderness at first frustrated physically the attempts of early and later pioneers to find a second Garden of Eden in the west, and it 'acquired significance as a dark and sinister symbol'. The pioneers therefore shared the Western tradition of 'imagining wild country as a moral vacuum, a cursed and chaotic wasteland'. So frontiersmen 'in the name of nation, race and God' saw themselves as civilising the New World by destroying wilderness and transforming it into cultivated landscape. This became 'the reward for his sacrifices, the definition of his achievement and the source of his pride' (Nash 1974 pp24-5). Such a frontier ethic is still strongly extant in areas like the Front Range of the Rockies and the neighbouring prairies. Here the presumption is with the developer, town planners have a minimal advisory role, and 'free' enterprise, civic pride and resolute independence are all invoked in the battle against the ecocentric environmentalist who pleads for preservation of wilderness.

As we shall see in section 3.5, in this struggle the ecocentric very much restates the ideas of the transcendentalists. They, in turn, took on the European romantic's ideas. Many of the latter, such as Chateaubriand, de Tocqueville and Byron, visited the US and sang the praises of its wilderness. These were echoed by a gradual perception among frontiersmen such as Daniel Boone that wilderness had aesthetic values. In the 19th century growing concern over the loss of wilderness 'necessarily preceded the first calls for its protection' (Nash p96) by such as the ornithologist J.J. Audubon, who had many occasions on his travels to observe the destruction of the forest. These calls led eventually to the National Parks movement which was inspired by Henry David Thoreau's phrase 'In wildness is the preservation of the world'. John Muir, who in 1892 founded the Sierra Club (which spawned the breakaway Friends of the Earth in 1969) was the guiding force behind the establishment of Yosemite National Park. He fought to set up a system of national forests, and resisted the infamous scheme to drown Yosemite's Hetch Hetchy Valley. He drew much of his inspiration from Thoreau (Opie 1971).

Thoreau has been epitomised as representing the values of the isolated individual living in nature and free of all social attachments. As an experiment he cast himself away from society for two years, and taking virtually nothing with him from the civilised world, he built a cabin and won a living from nature by the side of Walden pond in Massachusetts. His account of the experiment, written in 1854, rejected the materialism which saw only monetary worth in nature:

> I respect not his labors, his farm where everything has its price, who would carry the landscape, who would carry his God to market, if he could get anything for him ... on whose farm nothing grows free ... whose fruits are not ripe for him till they are turned to dollars. Give me poverty that enjoys true wealth (Thoreau 1974 p145).

Thoreau here shows bioethical sentiments - that nature is worth respect and reverence in its own right, regardless of its economic use to man. These are also evident in Ralph Waldo Emerson, whom Opie calls 'the leading exponent of transcendentalism'. In his first book, *Nature* (1836), Emerson wrote 'Such is the constitution of all things ... that the primary forms, as the sky, the mountain, the tree, the animal, give us a delight *in and for themselves*'. And, as with some later ecocentrics, one sometimes gets the impression that so strong is the integrity and purity of wild nature that man sullies it by his very presence. 'This pond [Walden] has rarely been profaned by a boat, for there is little in it to tempt a fisherman', said Thoreau. Certainly there is a sense in 'pure' preservationist transcendentalism that economic activity vandalises nature. 'This Sierra Reserve ... is worth the most thoughtful care of the government for its own sake ... Yet it gets no care at all ... lumbermen are allowed to spoil it at their will, and sheep in unaccountable ravenous hordes to trample it and devour every green leaf within reach ...' (Muir 1898). This purist stance was challenged by Gifford Pinchot, whose close affinity with Muir became strained as he developed his 'managerial' conservationist (technocentric) philosophy. 'The first great fact about conservation', wrote Pinchot (1910):

> is that it stands for development ... Conservation does mean provision for the future but it means also and first of all the recognition of the right of the present generation to the fullest necessary use of all the resources with which this country is so abundantly blessed.

As Opie says, this point of view 'still dominates today's federal land use programs although the ecology controversy in the 1960s has encouraged the "wilderness-for-wilderness-sake preservationists"' (the ecocentrics).

As well as bioethic philosophy, the transcendentalists also strongly embraced the views that contact with wild nature purifies and refreshes men and women, that people can find spiritual fulfilment in nature (that is, they can transcend their material existence via quasi-mystical experiences) and that all wild nature is a manifestation of God (pantheism). Thus Thoreau wrote that 'we need the tonic of wilderness ... we can never have enough of nature', and by Walden's banks he was 'affected as if in a peculiar sense I stood in the laboratory of the Artist who made me'. Emerson, in *Nature*, called the woods 'these plantations of God'; in them he saw all 'the currents of the Universal Being circulate through me. I am part or parcel of God ... The greatest delights which fields and woods minister is the suggestion of an occult relation between man and the vegetable'. Emerson's reference to the currents of a Universal Being strongly evokes one of the key ideas of the medieval Chain of Being, and later he also restates the metaphor of nature as a book: 'What is a farm but a mute gospel? The chaff and the wheat, weeds and plants, blight, rain, insects, sun - it is a sacred emblem ...'.

Muir, who called the Grand Canyon 'this grandest of God's terrestrial cities', strongly promoted wilderness' tonic and spiritual values:

> Thousands of tired, nerve-shaken, over-civilised people are beginning to find out that going to the mountains is going home; that wildness is a necessity and that mountain parks and reservations are useful not only as fountains of timber and irrigating rivers but as fountains of life. Awakening from the stupefying effects of the vice of over-industry and the deadly apathy of luxury they are trying as best they can to mix and enrich their own little ongoings with those of Nature, and to get rid of rust and disease ... some are washing off sins and cobweb cares of the devil's spinning in all-day storms on mountains (Muir 1898).

In these writings, too, one can see other ideas that commend themselves to modern ecocentrics. For instance the concept that there is more to nature than the sum of its parts is evoked by Emerson's sentence 'There is a property in the horizon which no man has but he whose eye can integrate all the parts, that is, the poet'. We should note that the integrator is not the 'scientific' systems ecologist, but the poet - the artist. Nonetheless, the empirical scientist would have to approve of Thoreau's detailed descriptions of the form and movement of the lobate sands which made up the banks of Walden pond. As Erisman (1973) says, 'for all its intuitive insights and mystical meditations, the *Walden* journal abounds in raw bold data'. Thoreau strongly conveys the idea that there is pleasure and reward in the close and systematic observation of nature. Detailed observation will also reveal the endless variety, complexity and fecundity of nature - its plenitude (see 3.1) - and this is a quality in which the transcendentalists revelled. 'I love to see that nature is so rife with life that myriads can be afforded to be sacrificed and offered to prey on one another', said Thoreau, while Emerson spoke of the 'unity of Nature - the unity in variety'.

3.4 THE COUNTRY AND THE CITY

We have noted that the romantics revolted against the 'excrescences' of industrial capitalism, such as vulgarity, poverty, squalor, materialism, pollution and ugliness. These excrescences were symbolised in the 19th-century city, and anti-urbanism is a major feature of romantic thought. Just as the romantics were part of a major attitudinal change towards the country, so, too, did they reflect a reversed perception of the city. As Tuan (1974) shows, the design of the ancient and medieval city often carried religious symbolism - it was a shrine to God as well as being an expression of man's highest cultural and technological achievements. This 'sacredness' contrasted with the 'profanity' of the wilderness which we have noted. With the increased importance of the industrial manufacturing function of the city, however, these positions were reversed, and just as the wilderness became sacred, so the city was regarded - especially by romantics - as a profanity (Tuan 1971B).

Thus Emerson wrote of how 'The tradesman, the attorney, comes out of the din and craft of the street and sees the sky and the woods, and is a man again', while the poet nourished by nature will not lose the benefit altogether 'in the roar of cities or the broil of politics'. Thoreau also shows anti-urbanism when he maintains that 'in nature, not in the pomp and parade of the town, the individual may walk with the Builder of the universe'.

Once again, the American romantics were here echoing a series of attitudinal changes which had taken place in Europe. 'In Renaissance times the city had been synonymous with civility, the country with rusticity and boorishness ... Yet long before 1802, it had become commonplace to maintain that the countryside was more beautiful than the town' says Thomas (1983 pp243-4), who notes that this reversal was connected with the deterioration of urban environments. Dirt-laden London air had been the subject of complaints in the 13th, and again in the 16th, century, while in the 17th and 18th centuries Oxford, Newcastle and Sheffield also experienced atmospheric pollution. Thomas asserts, however, that contemporary thought was more preoccupied with the city's immorality than its dirtiness. He quotes John Norris, who wrote:

Their manners are polluted like the air,
From both unwholesome vapours rise
And blacken with ungrateful steams the neighbouring skies.

In noting these sentiments as part of an attitudinal change associated with romanticism, we must be careful to recognise that we are once again identifying the resurgence of one strand of a dualism which (as with the rational-romantic dualism itself) can be traced back very far. As Williams (1975 p9) put it:

In the country has gathered the idea of a natural way of life: of peace, innocence, simple virtue. In the city has gathered the idea of an achieved centre of learning, communication, light. Powerful hostile associations have also developed: on the city as a place of noise, worldliness and ambition: on the country as a place of backwardness, ignorance, limitation. A contrast between city and country, as fundamental ways of life, reaches back into Classical times.

From time to time in cultural history one aspect of this dualism surfaces while the other becomes relatively dormant, but the two strands are always there, in fundamental tension. In fact, Williams shows that there is also a powerful tendency in all of us in the West to associate the country's more benign image with the *past* - a past in which things were invariably better than they are now. In this tendency we emulate not just the romantics but those writers from the classical poets onwards who have always promulgated the fiction of Arcadia - of a countryside of simplicity, virtue and man-nature harmony. This is a pastoral idyll. As Leo Marx says, it is a [senescent] vision of a 'Golden Age, of grazing flocks, unruffled waters and a calm, luminous sky, images of perfect harmony between man and nature ...'. And of course it is substantially a *myth*. For one thing, it has no place in it for the necessity of human labour. 'In the pastoral economy nature supplies most of the herdsman's needs and, even better, nature does virtually all of the work' (Marx 1973). In some romantic paintings this 'magical extraction of the curse of labour' was achieved by simply removing any labourers from pastoral images:

> The actual men and women who rear the animals and drive them to the house and kill them and prepare them for meat... these are not present: their work is all done for them by a natural order (Williams 1975 p45).

The 'Picturesque' movement's countryside excluded not only agricultural workers but also the ploughed field and in general the works of man, which were considered distasteful. In this way the patrons of such art were not reminded of the more baleful aspects of the production from which their wealth came.

Nonetheless labour was an important, if unrecognised, part of the pastoral myth, because in the myth what constituted 'nature' was not in fact wilderness. It was, rather, a 'garden' - wilderness after it had been tamed and cultivated. It was a middle landscape between city and wilderness (Tuan 1971B) - a 'civilised' kind of nature which was revered and regarded as 'true' nature. This tradition of seeing idealised nature as a garden is found not only in the Arcadian myth, but also in the Christian religion, where the Garden of Eden is associated with God's grace. When Adam and Eve fell from grace they had to leave the Garden. But the true Christian, according to William Blake, must labour to regain the garden and strive against the evil forces of industry and materialism:

> And did those feet in ancient time
> Walk upon England's mountains green?

> And was the holy Lamb of God
> On England's pleasant pastures seen?...
> And was Jerusalem builded here
> Among these dark Satanic mills?

It will be remembered that we must be prepared to fight both mentally and physically to replace Satan's excrescences by a holy city in 'England's green and pleasant land'.

Such a fight was taking place in North America in the 18th century, where the pioneer cultivators were labouring to create a middle landscape. Their importance was asserted by Thomas Jefferson, when he said 'Those who labour in the earth are the chosen people of God... Corruption of morals in the mass of cultivators is a phenomenon of which no age or nation has furnished an example'. And he appears to have followed the romantics when he went on to describe the urban workers: 'The mobs of great cities add just so much to the support of pure government, as sores do to the strength of the human body': quoted in Marx (1973).

Here, environmental images of city and country were having social images added to them. It was this potent imagery which also gave the 19th-century garden city movement its impetus. It was summed up by Ebenezer Howard (1902) in his attempt to resolve that tension between country and city which romantics dramatised:

> There are in reality not only... two alternatives; town life and country life, but a third alternative, in which all the advantages of the most energetic and active town life, with all the beauty and delight of the country may be secured in perfect combination. And the certainty of being able to live this life will be the magnet which will produce the effect for which we are all striving: the spontaneous movement of the people from the crowded and unhealthy cities to the bosom of our kindly mother earth.

3.5 THE ROMANTIC LEGACY IN MODERN ECOCENTRISM

In the 20th century this spontaneous movement has taken place, though not generally into garden cities. Rather, people have sought to escape the city by creating suburbs, or, if more affluent, by living in the country villages surrounding towns. In so doing, the latter have gained an adventitious population which has helped to transform the face of the countryside. But at the same time these newcomers have been in the vanguard of the environmentalist movement, giving their wealth, time and articulateness to groups such as the Council for the *Protection* (emphasis added) of Rural England and the National Trust and the anti-airport groups (Lowe and Goyder 1983, Pepper 1980). They have constituted the more conservative wing of ecocentrism, and their motivation in wanting to acquire access to and to protect what Shoard (1980) has called the 'traditional' landscapes of England (in reality man-made middle landscapes dating from the enclosures) seems to have been derived substantially from preconceived images of the benefits of rural life and of what 'true' country is, or, rather, *should* be. For the images constitute a form of rural 'nostalgia', as Harrison (1982A) puts it, on the part of people whose original home was *urban*, or *suburban*, not in the country. They do not know the reality of the latter as a place of production. So the images are very romantic. Newby writes that to most inhabitants of urban England the countryside 'Supports a serene idyllic existence enjoyed by blameless Arcadians, happy in their communion with nature'. It involves the 'primitive values' of rural life, the 'Good Life' with a vengeance. He continues that this has been part of a broader romantic resurgence, despite or in response to, the harsh economic realities of 1970s England. We have created:

> a peaceful, if mythical, rural idyll out beyond the high rise flats and the Chinese takeaways which, if not quite inhabited by merrie rustics, is at least populated by a race which is supposedly attuned to verities more eternal than the floating pound and the balance-of-payments crisis... Somewhere... at the far end of the M4 or the A12, there are 'real' country folk living in the midst of 'real' English countryside (Newby 1979 pp13-14).

This is an escape from an urban culture that we cannot come to terms with - into a world where we can find a tonic for our psyches and emotions. It is based upon sentiments encapsulated by Nicholson's (1970) rather sweeping ecocentric/romantic rejection of the city:

> Modern man has improvised without foresight a street-and-cement wilderness about him which is in many ways more frightening and uncongenial to him than the natural wilderness from which he sprang. Do people really want this great urban sprawl, or does it alienate and unsettle them?... the megalopolitans of the world are an increasingly depressed and frustrated human group.

If we make our escape to the country, we try, through environmentalist groups, to make our rural myths real. Hence the clash between conservationists and modern farming methods, urban encroachment, motorways, airports and 'development' in general is a clash to 'preserve' what has in a sense never really existed.

The harshness of down-to-earth materialism of the 'real' country is only occasionally brought to our attention by the media (e.g. through the book and film of Blyth's (1969) *Akenfield*). In the main we find them happy to feed the romantic notions - through Laura Ashley fabric designs, 'country fresh' farmhouse soup, cheese and other 'natural' products, through 'everyday stories of country folk' on radio, and through dramatisations of James Herriot and R.F. Delderfield on TV. Of the latter's work, Giddings (1978 pp588-9) writes that not only is the rural myth strong in it:

> it is almost totally inaccurate... There is something wholesome about agricultural work in the open air - even if we only watch it on our colour TV... *A Horseman Riding By* is one example of a substitute art which protects us from seeing the realities.

The realities include steady declines in rural services, like transport, medical and educational provision, and exceedingly low pay for agricultural workers, and Newby and others believe that it is dangerous for us not to be aware of them. For, as Thomas (1983 p268) points out, 'the pull of wild nature can always be recognised as an essentially anti-social emotion'. Presumably he refers to the escapism of this pull, which allows us to forget the need to address the problems that face the mass of people - urban and rural underprivileged alike - who cannot, like us, get away from them.

Escapism, as well as the more positive romantic desire for spiritual regeneration, was what partly inspired the 'wilderness movement' in Britain and America that led to the eventual establishment of the National Parks. Shoard (1982) describes this move in Britain, and she notes that no Parks were created in the lowlands. This was partly because of a 'romantic legacy' which helped to shape the perceptions of key figures in the conservationist movement which was pressing for the Parks. Vaughan Cornish, for example, 'a typical child of the romantic movement', thought that mountains and sea-lapped headlands constituted the supreme form of landscape. And John Dower, whose 1945 government-commissioned report on National Parks was most influential in determining what countryside was to be included in them, was essentially an 'upland man' - raised in Northumberland and the Yorkshire Dales.

Shoard has discovered, through in-depth interviews with moorlands devotees, what are currently thought to be the attractive features of these upland environments. They are nearly all qualities of the wilderness which the romantics also eulogised. There is *wildness*, the antithesis of domestication, and *'naturalness'*, an apparent absence of human handiwork (even though the devotees know full well that moorlands were created by man 4000 years ago). There is their height, perhaps fulfilling people's needs for aloofness, and the *freedom* to wander at will, both satisfying the desire for *solitude* and *individuality*. And there is that much-praised quality, *simplicity*, in the homogeneity of form. The openness and grand vistas which are afforded, in which the sky dominates, facilitate *communion with the creator*, providing 'almost a a religious experience'.

Another factor in the choice of upland moorland was the precedent set by North America, where National Parks were established in the 1870s principally to preserve spectacular and wild landscape. The Parks lobby is still a strong political force, and its campaigning hinges upon that strongly romantic sentiment enunciated by John Muir, the bioethic. O'Riordan describes how in 1972 the American Conservation Foundation proposed a 'Parks first' policy to preserve wilderness - not primarily for tourism but for nature itself. Where the two interests clashed, it was proposed that public access should defer to the 'rights' of nature. And American lawyers have attempted 'to bestow legal rights of existence on animate and inanimate objects along the lines advocated by Leopold (1949)'. Supreme Court Justice Douglas is reported to have said, in a dissenting opinion where judgement had been given against the Sierra Club's attempt to prevent increased access to some National Forest:

> Those who like it [the Forest], fish it, hunt it, camp it, or frequent it merely to sit in solitude and wonderment are legitimate spokesmen for it... Those who have that intimate relation with the inanimate object about to be injured, polluted or otherwise despoiled are its legitimate spokesmen (O'Riordan 1981A pp5, 280).

The extension of this would be a 'guardianship theory whereby citizens could sue on behalf of natural objects'. Already, in laws protecting endangered species there is the idea that they have legal rights which can be asserted for them by any concerned citizen. And ecocentrics call for more protection:

> ... the world needs a new Bill of Rights: a Bill of Rights for all future generations of man, but also a Bill of Rights for the four million other sorts of plants and animals with which we share this planet earth. A Bill of Rights for nature, no less (Bellamy 1983).

Explicit in this call is the romantic, bioethical idea that 'We share earth with other species. Some we cannot do without and for others we should have a sense of responsibility... Man has trusteeship for nature' (Allsopp 1972 p91). There is a strong pragmatism here, but there is also unmistakably the assertion that man should be the steward even of those parts of nature for which he has no obvious use or need.

Underlying this is the strong sense of *interconnectedness*, of *wholeness* which characterises ecocentrism, e.g. in Commoner's 'spaceship earth' concept, in Capra's 'organic systems view' (see Chapter 4), or in the inspiration for the *Whole Earth Catalog* (Brand 1971). Erisman (1973) draws a direct link between this aspect of ecocentrism and Thoreau's observation that 'The pure Walden water is mingled with the sacred water of the Ganges... melts in the tropic gales of the Indian seas, and is landed in ports of which Alexander only heard the names'. This kind of complexity of *natural* interrelationship is acceptable to ecocentrics, whereas *man-made* complexity is to be eschewed in favour of the romantic simplicity, that is, contact with 'simple' nature. This is a rather paradoxical position. Erisman quotes 'Ecology Action East's' manifesto (1970), which asserts 'the right to experience on a daily basis, the beauty of the natural world, the open, unmediated sensuous pleasure that humans can give to each other, the growing preference for the world of life'. Here is a call for 'unmediated' i.e. unsophisticated, simple, basic, enjoyment of nature, including human nature.

Erisman also identifies in environmentalism that neo-romantic sentiment which underlies all the others - the *anti-material*, anti-scientific, anti-rational. He cites Commoner's (1967) arguments against using modern technology as broadly resembling 'the anti-technology arguments advanced by virtually all of the New England Transcendentalists'. Though the latter seemingly concentrate on the spiritual rather than environmental effects of technology - as in 'We are in great haste to construct a magnetic telegraph from Maine to Texas; but Maine and Texas, it may be, have nothing important to communicate' (Thoreau 1974 p48) - Erisman thinks that in reality the spiritual is equally important for environmentalists. He quotes Reich's (1970) lament: 'In our country they [technology and production] pulverise everything in their path'. 'Everything' includes environment, history, tradition, social structures, privacy and beauty. Erisman concludes that modern American environmentalists are 'not the *avant garde* of a new, but only the most recent manifestations of an old, romantic tradition'.

Not that many of these affluent and educated neo-romantics would be sorry to be regarded as in the forefront of a new order. The fashionable 'concerned environmentalist set', with their open-toed sandals, macramé, muesli, therapy, and consciousness-raising are mercilessly exposed in McFadden's (1977) hilarious satire on 'A year in the life of Marin County', and in Bookchin's (1980) more serious denunciation of the American predilection for instant gratification, whether it be in fast food or fast philosophy.

This criticism conveys a sense that it is not merely harmless fun, and that too much neo-romanticism is a bad thing - escapist, as we have suggested, and also superficial and hypocritical. As a civil rights worker reputedly said after Nixon's assertion, in his 1970 State of the Union Address, that *the* question for the 1970s was the environment, 'It is a sick society that can beat and murder black people on the streets, butcher thousands in Vietnam, spend millions on arms to destroy mankind, and then come to the conclusion that pollution is America's number one problem'.

Erisman, too, draws attention to a 'potentially tragic irony' behind the 'new/now movement - those intensely romantic Americans' who strive like Emerson and Thoreau 'to break the ties of rationality and materialism and create a world of emotions and individual spontaneity'. For he thinks that there should in reality be a balance between romanticism and rationalism:

> in the disciplined part of Thoreau's work, rather than in his wildness, may lie the preservation of the world. Without a foundation of systematic scientific knowledge, ecological speculation becomes little more than a simplistic attack upon capitalism, the Protestant Ethic, and the entire Judaeo-Christian tradition.

The next chapter intends to show that there is, indeed, also a foundation of systematic scientific knowledge and philosophies upon which modern ecocentrism rests.

CHAPTER 4

THE SCIENTIFIC ROOTS OF ECOLOGICAL ENVIRONMENTALISM

4.1 MALTHUS AND NEO-MALTHUSIANS

4.1.1 Heralding the Ecocentric/Technocentric Debate

Just as we can fairly readily identify elements of 19th-century romanticism in modern ecocentrism, so, too, can we see much of the influence of Thomas Malthus (1766-1834) in the modern debate about limits to growth. We could go further back in history in tracing the roots of concern about a possible conflict between population growth and subsistence. But, as Hollingsworth (1973) says, 'It was the attempt to be specific and mathematical about the dilemma of an expanding population that is Malthus' special contribution to thought'. In other words, Malthus' *Essay on the Principle of Population* can be seen as an attempt at making a *scientific* contribution to the debate. While it is easy to identify romanticism's input into ecocentrism as *non*-scientific, the notion of Malthus as science is slightly less accessible to sophisticated modern students of science. Hence we shall have to examine the idea a little more closely.

First, however, it is worth dwelling on the background of ideas of the time in which Malthus wrote, because at the end of the 18th century industrial capitalism bolstered by scientific and technological advances was reaching a new prominence. A debate about how far it could take society along the road of progress was already under way. It was a debate which preempted the modern one between ecocentrics and technocentrics, and it held within it elements of the most fundamental philosophical question about how much man is a creature of free will as against being at the mercy of external forces which he cannot control.

In this debate many see Malthus as 'explicitly and militantly pessimistic' (Pavitt 1973). Against a background of buoyant optimism, in which faith in human progress was becoming equated with constant scientific and technological advance, Malthus proposed the idea that the earth sets limits to population growth and human wellbeing. Glacken (1967) believes that the roots of this idea lie in contemporary observations by travellers and explorers of just how teeming the earth was with a rich variety of life.

This abundance and fecundity among living organisms - part of earth's plenitude (see 3.1) - was so great as to suggest that each species could multiply indefinitely were it not for the existence of checks to such multiplication. It led people towards what Glacken calls some 'absurd extravagances', such as Count Buffon's observation in 1749 that a single species could cover the entire earth were no obstacles placed in its way, or Darwin's view that the elephant (the slowest breeder) could stock the whole earth in a few thousand years - or Malthus' own statement in his *Principles of Political Economy* that the human race could theoretically fill not only earth but all the planets in the solar system. It was clear, however, that such extravagances had not come about, so therefore populations were being held in check by wars, plagues, famines and other 'unhealthful events', and by competition within and between species.

Indeed, some thought that such events had been only too effective, to the extent that the number of people on earth had actually declined since the times of classical Greece and Rome. This thesis, of natural *senescence* (see 3.4), or decay of the earth, is the opposite of the idea of progress. It involves, says Glacken, the organic analogy (see 2.2.1 and 3.1). Mills points out that this analogy is essentially a pessimistic one, because if the comparison holds good, then the earth must be decaying and dying, just as all organisms do from the day of their birth:

> You are surprised that the world is losing its grip? That the world is grown old? Think of a man: he is born, he grows up, he becomes old. Old age has its many complaints: coughing, shaking, failing eyesight, anxiety, terrible tiredness. A man grows old; he is full of complaints. The world is old; it is full of pressing tribulations. (St Augustine 410 AD, quoted by Mills 1982 p243).

Writing in 1755, Robert Wallace postulated that the ancients had been more numerous because of their moral superiority over the moderns. His thesis was essentially that the environment sets limits to the numbers and the wellbeing of man, and Malthus developed this into the view that there were insurmountable natural barriers to the achievement of the utopian dreams of those who believed in the possibility of unrelenting social progress through institutional reforms.

In advancing it, Malthus stood in particular opposition to Godwin and Condorcet. The latter, in a *Sketch for a Historical Picture of the Progress of the Human Mind* (1793) wrote that the 'perfectibility of man is truly infinite', and that there was no limit to progress other than the duration of the globe. Both felt that 'overpopulation' in relation to subsistence needs would not constitute a serious threat to progress, for if such a contingency arose in the distant future it could be overcome with the aid of technological skill, or by the wisdom of future societies who would see the danger and refrain from overbreeding. As has been suggested, such optimistic assertions corresponded to the *zeitgeist*, and they were associated with the reality of rising revolutionary fervour in Europe.

It was to counter such revolution that Malthus preached in the first edition (1798) of his *Essay*, which was essentially a polemic against Godwin and Condorcet and revolutionaries like Tom Paine (*The Rights of Man,* 1791/2). It consisted of what became Books III and IV of later editions of the work, and it did not have the 'scientific data' which appeared in the 1803 (2nd) and subsequent editions as Books I and II. In eventually producing such data, Malthus showed himself also to be at least partly a man of his time. For, as Petersen (1979) points out, the age was one where moral philosophy (the direct application of Christian values to current issues) was giving way to political economy - the defence of 'natural' rights set in the context of a 'natural' order of things with less and less reference to a deity. In other words, it was a period where the empirically-derived laws of nature (and the classical economic laws being formulated by Adam Smith) were attaining ascendancy over belief in the supernatural - an 'enlightenment', with a break in thinking from religious to secular views. The establishment of such laws required the collection of data. Because he talks of natural rather than Christian laws, Malthus was seen by some as an atheist - especially as he seemed to be warning against the injunction in Genesis to 'be fruitful and multiply'. However, we need to distinguish here between *primary* and *secondary* causes. Malthus held the laws of nature as secondary causes or influences on man. The primary cause, or ultimate determinant on man, was God. He worked via the medium of natural laws.

4.1.2 Malthus' Thesis

Malthus' views on population hinge upon his assertion that population growth - unless it is checked - will tend to outstrip increases in the level of food supply. While population:

> when unchecked, goes on doubling itself every twenty-five years, or increases in a geometrical ratio...the food to support the increase...will by no means be obtained with the same facility. Man is necessarily confined in room when acre has been added to acre till all the fertile land is occupied, the yearly increase of food must depend on the melioration of the land already in possession. This is a fund, which, from the nature of all soils, instead of increasing must be gradually diminishing. But population, could it be supplied with food, would go on with unexhausted vigour; and the increase of one period would furnish the power of a greater increase of the next, and this without any limit (Malthus, 7th ed. 1872, Book 1, Chapter 1).

This tendency for population to push against the bounds of food supply is due to the passion between the sexes, and it is inescapable, though it can be contained. Agricultural production, at best, can increase only in arithmetic proportions... 'the means of subsistence, under circumstances most favourable to human industry could not possibly be made to increase faster than in an arithmetical ratio'. Thus, assuming that the best increases in agricultural production can be secured, 'the human species would increase as the numbers 1,2,4,8,16,32,64,124,256, and subsistence as 1,2,3,4,5,6,7,8,9. In two centuries the population would be to the means of subsistence as 256 to 9; in three centuries as 4096 to 13, and in two thousand years the difference would be incalculable'.

Malthus was sceptical that agricultural production could be indefinitely increased, even on this theoretical scale, because of the increasing need, referred to above, to use marginal land. Obviously, this 'principle' or law of population represents an impediment to 'the progress of mankind towards happiness'. However, while in plants and irrational animals potential population increases are reached, to be repressed afterwards by want of room or nourishment, humans, by process of reason, react in a more complicated way to 'the constant tendency in all animated life to increase beyond the nourishment prepared for it'. Impelled by instinct to increase, he asks if inability to provide for his children will not bring him down economically and make him obliged to charity. This reasoning prevents early marriage, and such 'moral restraint' forms one of a series of *preventive checks* to unbounded population increase. Although chaste restraint was approved of, some other preventive checks could constitute 'vice' (e.g. contraception, abortion, promiscuity with prostitutes, homosexuality). Malthus thought he observed that preventive checks were applied unevenly - more so in European than in 'primitive societies' where control over the food supply is slight and therefore population tends to oscillate around its means of subsistence (as with an animal population). The preventive checks were also deemed less applicable to the poor and uneducated classes in Europe, whereas wealthier and better educated people have later marriage and smaller families. So for the poor any increase in income would lead to increased family size, because they would not practise preventive checks and because other restraints on procreation such as starvation, disease and illness would be ameliorated. These kinds of restraints, together with wars, were collectively known as *positive checks* on population growth, or 'misery'. Where preventive checks were not applied, then positive checks would come more strongly into play - unless poor relief interfered with them.

Hollingsworth (1973) voices some typical objections to the geometrical/arithmetical theory. While, he says, it can be accepted that human populations are biological and *tend* to increase geometrically, 'human ingenuity is not subject to an arithmetic law of growth'. Malthus has underestimated technological potential and, in any case, there is no particular basis for an arithmetic progression in food increase rather than any other.

However it is against the economic stage of Malthus' thesis that technical demographic criticisms are particularly commonly levelled. This stage, according to Pavitt (1973) carried the assumption - based on differences between Europe and the Americas - that rates of population growth for some groups *increased* steadily with rising income. Economic growth, particularly if it reached the poor, would lead to increased population which would defeat the ends of economic growth because the wealth would have to be shared among more people. Glacken interprets this as a cycle which tends to repeat itself and leads to pulsed rather than steady population growth. Population increase leads to increased 'misery', including lower wages (when there are more people the labour market is depressed). Lower wages induce cultivators and landowners to employ more labour, thus agricultural productivity rises, increasing carrying capacity and encouraging a further population surge. This will go on until the ultimate limits are reached. Thus, during depression poverty leads to postponement of procreation (of 'marriage'). But with the return of prosperity more people will begin families, building up population pressure on subsistence. Since this positive correlation between agricultural production and the number of marriages (and children), thought Malthus, applied mainly to the lower classes, it was hidden from demographers and historians, because most recounted history tended to be the history of the upper classes. Pavitt points out that this analysis has been absorbed into the main current of economic thought, even though it is incorrect. For with economic growth in the West (and some redistribution) has come increased contraception and *lower* population growth. Chase (1980) says that Malthus did not pause to explain *why* rising wages would trigger a rise in births, and both he and Simon (1981) point out that modern demographers have coined the term 'demographic transition' to describe the observed fact that as a nation's family income and living standards rise and death rates fall, birth rates also decline.

Although there are those (e.g. Schnaiberg (1980) and Petersen) who think that these criticisms are misinformed, and who contend that Malthus developed over time into an optimist about the prospects of limiting population through prudential restraint, it does not appear that Malthus changed very much his views on the social implications of his population principle. They seem to have remained unaltered into the last edition of the *Essay*. They were that the 'middle and higher classes of society' should direct in a *selective* way their efforts to improve the condition of the poor. Although the rich could do much by advising and instruction, 'encouraging habits of prudence and cleanliness', through dispensing 'discriminate charity' and through any other means leading to strengthened preventive checks, they should not make early marriages more generally possible (e.g. by improving wages and living conditions on a large scale, through supporting Poor Law relief). For large scale wealth redistribution would simply, through the natural laws of population, encourage the lower classes to increase their numbers. It would, as Petersen describes it, 'mean less that people eat well but that more eat', increasing and not decreasing misery. This 'important truth...flowing from the principle of population' means that large scale poor relief would be futile and self-defeating, and that 'the rich do not in reality possess the *power* of finding employment and maintenance for the poor' (7th ed, Book 4, Chapter 14).

4.1.3 Malthus as a Scientist
Earlier considerations on the nature of science (see 2.5) enable us to accept Chase's definition of science as 'Any branch of systemized knowledge of nature and the physical world in which information and concepts are derived from observation, analysis and experimentation in order to determine the nature or principle of what is being studied'. Bearing this in mind we should be able to examine critically the claim

that Malthus' population work was 'scientific', and that the 'principle' of population was a scientific law. We shall be looking to see a *'general statement* of fact, methodologically established by *induction* on the basis of *observation* and experiment and usually... expressed in a *mathematical* form... Ideally, scientific laws are strictly *universal* or deterministic in form, asserting something about all members of a certain class of things...' (Bullock and Stallybrass 1977, emphases added).

Modern biographers and critics do not seem to doubt that Malthus can be viewed as a scientist. Hollingsworth, a demographer, describes him as a moralist and politician, but also as a social scientist (sociolgist and economist) and a demographer 'tracing the population trends in various countries as accurately as he could, and arguing about the reasons for their differences'. Schnaiberg calls Malthus an 'early and distinguished' social scientist, and Malthus seems to have achieved some distinction among his contemporary scientific community, being an FRS, a Fellow of the Royal Statistical Society, a member of the Political Economy Club, and other eminent bodies.

His *Essay* appears to have replaced deduction by induction as it developed. The first, polemical, edition started with two postulates of 'fixed laws of nature' - that 'food is necessary to the existence of man' and that 'passion between the sexes is necessary and will remain nearly in its present state'. From these it moved to the eventual deduction that the power of the population to increase was greater than that of earth to provide food, introducing geometrical and arithmetic ratios as additional postulates.

The approach, however, changed for the second and subsequent editions, when Malthus first laid out empirically-derived evidence from which he induced the principle of population. Petersen describes the first essay as a mainly deductive book of some 55000 words, while the second was a 'new book' in which the expansion of theory and of illustrative data increased the work four-fold. No less an economist than Alfred Marshal rated the latter as 'One of the most crushing answers that patient and hardworking science has ever given to the reckless assertions of its adversaries' (quoted in Petersen 1979 p53). Hollingsworth described Books 1 and 2 as devoted to a remarkably complete demographic survey for the time, and Petersen thought 'Malthus spent a lifetime labouring to improve the statement of his theory and gathering facts on which to base these emendations' (idem p49) though he admits that some thought the demographic data 'of dubious quality'.

Indeed, the survey seems very complete, so that any law which derives from it can be thought of as universal. Book 1 was about the checks to population in 'the less civilized parts of the world' and in past times. It covered the American Indians, South Sea Islands, ancient Northern Europe, modern 'pastoral nations' (Asia), parts of Africa, Siberia, the Turkish dominions and Persia, Indostan and Tibet, China and Japan, and the ancient Greeks and Romans. Book 2, on the checks to population in modern Europe, covered Scandinavia, Russia, Middle Europe, Switzerland, France and the UK, and it ended with a chapter of general deductions.

The sources of this evidence, when population data was generally sparse and inaccessible, were various, and mainly secondary. There were reports from missionaries, voyagers and explorers, there were geographies and histories and books of observations on customs and habits of far-off lands. There were also population registers and indices of prices. James (1979) says that 102 authorites were consulted. In addition, informal material was collected assiduously from travellers - especially from a life-long friend, Edward Clarke, who went to Russia and many other places, seeking all the time answers to Malthus' written questions on checks to population. And there were Malthus' own travels to France and Switzerland, and to Norway in 1799. Petersen (1979 p50) paints a picture of him as an enthusiastic scientist, 'earnest without being pedantic; interested in just about everything', and making jottings on finance, economic trends, profits and wages and population - including the institutional arrangements to prevent early marriage.

Peterson admits, however, that his Scandinavian trip was 'a rather casual affair', that he knew no Norwegian and that he visited only portions of the country. Certainly we might think his evidence a little unsystematic and anecdotal. James (p97) records how his chapter on Switzerland used the 'Philosophic discourse of the driver of our char on the overpopulation of his country, he [the driver] complained much of the extreme early marriages, which he said was the "vice du pays"' (extract from Harriet Eckersall's diary). And Chapter 4, Book 1, on the checks to population among the American Indians is typical of the long rambling accounts with their tenuous sweeping generalisations which make it difficult for modern readers to see his strengths as a scientist (emphases added):

> It was *generally remarked* that American women were far from being prolific. This unfruitfulness *has been attributed by some* to a want of ardour in the men towards their women... it... *probably* exists in a great degree among all barbarous nations whose food is poor and insufficient, and who live in constant apprehension of being pressed by famine or by an enemy. Bruce *frequently takes notice of* it, particularly in reference to the Galla and Shangalla, savage nations on the borders of Abyssinia, and Vaillant *mentions* the phlegmatic temperament of the Hottentots as the chief reason of their thin population. It *seems to be* generated by the hardships and dangers of savage life...'

So it goes on, and on - with stories of habits and customs in marriage, of the kinds of common diseases in different countries, of cannibalism (positive or preventive check?), of propensity to wars, of agricultures and soils and climates. And, despite this attempt at empiricism, we cannot but agree with Hollingsworth when he says 'The real weakness of Malthus as a demographer is, to a modern mind, that he is too fond of general theories and not interested enough in empirical results'. This is startlingly supported in a letter, which James quotes, from Malthus to the Director of the Statistical Bureau in Vienna: 'It is quite true as you observe that the vast provinces belonging to the Austrian Empire have come very little under my notice. One reason was...that the statistical information relating to them was not very easily accessible; and another was, *that the information which I could collect without much difficulty appeared to me to embrace a sufficient number of instances to establish the principle which I had in view...*' (emphasis added). This extract supports Chase's contention that Malthus... 'was not a man to be intimidated by mere facts. To his dying day he was to cling to his central dogma...' This dogma was the stubbornly maintained proposition that the ratio between levels of human food and baby production was fixed by natural laws.

Certainly, Malthus, in his world survey, clearly maintained that he was seeking to establish a natural law - that is a universally applicable principle, operating uniformly in time and space. As Glacken remarks, the whole earth was considered the proper unit of study, and the principle of population was thought constant in its operation - not operating with more intensity in, for example, China or India than in Europe. As Petersen describes it, on the scale of the whole world and over all of history population constantly bears a regular proportion to the food the earth is made to produce. Hollingsworth attributes this desire to reduce 'human demographic and social development to two mathematical laws, of geometric and arithmetic progression' to Malthus's training. As a maths undergraduate at Cambridge 'he must have been fascinated by precise mathematical formulations of economic laws'. And in his work he included some tables showing the relationship he saw between the proportion of marriages to births and deaths in a population and population growth rates.

There remain questions, however, of how valid these relationships were, whether they were precisely mathematically structured or not. Although Malthus may have seen a positive correlation between economic growth and the number of marriages among the poor, to claim that the correlation evidences *causal* relationship is quite another matter. Pavitt believes that from the data available to him Malthus could have reached different interpretations, leading to different predictions about future events. It was true that at the time there was little empirical justification for associating economic growth with greater health and civilisation and declining population growth - indeed from the visual evidence it appeared that only dirty and overcrowded towns were associated with economic growth. But on the other hand Malthus did have data showing that healthier (and wealthier) communities with lower death rates also had lower birth rates. He could have induced a causal relationship between the two and gone on to make appropriate social recommendations involving major wealth redistribution. That he did not appear to advocate much change from the socio-political *status quo* is perhaps attributable to a certain lack of objectivity. Whether such a lack disqualified him from being regarded as a scientist, however, is unlikely, and is a theme which will be returned to (Chapter 5).

4.1.4 Neo-Malthusianism
We will consider two aspects of neo-Malthusianism, but will discuss only one of them in this chapter. This aspect concerns the extent to

which Malthus' thesis itself is advanced by any modern commentators - who would be regarded as 'neo-Malthusians' not only because of this agreement, but also through their use of similar (i.e. scientific) methods to arrive at it. The second aspect of neo-Malthusianism is the use of the term in an extended sense to apply to those who arrive at similar *deductions* to those of Malthus from the 'principle of population' about *social policy*. Such deductions are held to derive from *ideology*. Of course, a scientist's ideology is likely to influence his methods and the 'laws' he identifies in his work, so it is difficult to separate this from the conclusions which are drawn. (We have already discussed evidence of ideological bias in Malthus, even though we have tried to restrict ourselves only to considerations of his scientific method.) Nonetheless, we shall attempt to maintain this artificial separation for the present, and hold over until later the analysis of explicit Malthusian and neo-Malthusian ideology (see sections 5.3.2 and 7.3.3).

There is no doubt that some modern environmentalists are neo-Malthusian. Garrett Hardin made his allegiances quite clear in this ecocentric poem, which also contains the view that man of himself defiles nature (see 3.2.2):

Malthus! Thou should'st be living at this hour:
The world hath need of thee: getting and begetting,
We soil fair nature's beauty...
Confound ye those who set unfurled
Soft flags of good intentions, deaf to obdurate obstinacy'
(From Hardin 1964 p88-9, quoted by Chase 1980)

The 'obdurate obstinacy' is the *fact* of the ecological *law* of carrying capacity. This scientific law is held by ecologists such as Hardin to apply universally, to human as well as animal populations. It means that populations in a given ecosystem will increase exponentially (the equivalent of Malthus' geometric ratio) until limits set by food and other resources (increasing more slowly or not at all) are reached. These limits represent the carrying capacity of the ecosystem, and population numbers cannot rise beyond those dictated by the carrying capacity. At best they can remain in equilibrium with it, giving a flat S-shaped curve to describe the movement of population growth over time. Or, if 'biotic potential' (an excess of births over deaths) pushes the curve beyond carrying capacity then 'environmental resistance' will swiftly push it down again. In other words, there will be a rapid die-off, giving a 'J-shaped' population curve (the top of the J finishing on a downstroke, Kormondy 1969). Those with 'soft flags of good intentions' (the equivalent of Malthus' social revolutionaries like Tom Paine and believers in unending progress like Condorcet) are enabling biotic potential to increase thus ensuring future catastrophe in the form of a J-shaped curve - a die-off resulting from the inexorable operation of scientific law. Thus, as Burch and Pendell (1947) put it, human population numbers are inevitably limited by either conscious controls (preventive checks) or famine, disease and war (positive checks).

However, the neo-Malthusians like Hardin add a further dimension to the debate, arguing that another element of the 'misery' which results from overpopulation lies in deteriorating environmental quality through the pollution associated with excess economic activity generated by large numbers of people in search of high living standards. Thus in 'getting' as well as 'begetting', 'We soil fair nature's beauty'. Another way of putting this is to say that 'people pollute', which was the theme of population control movements in the early 1970s inspired by Hardin and Ehrlich (1970) - such as 'The Environmental Fund' and 'Zero Population Growth' (see 7.3.3).

This addition of a new physical constraint in the form of upset environmental/ecological balance is noted by Cole, Freeman, Jahoda and Pavitt (1973), whose critique of the *Limits to Growth* argues that the world-famous Forrester and Meadows study used the same techniques for shock effect as did Malthus. The *Limits*, too, projected exponential versus arithmetic growth well into the future. The models used were 'part of a revival of interest in the preoccupations of the classical economists'... and 'exactly the same parameters are explored' (with the addition of environmental degradation). 'In all cases growth is seen to be impossible because of physical limits, and whether and when "catastrophe" happens is rigidly determined by these limits' (Pavitt 1973).

Freeman (1973), in a critique of the *Limits* entitled 'Malthus with a computer', thinks that the whole study is not strictly Malthusian, but the approach to world forecasting is so. It is essentially 'scientific', being numerate and clearly formulated (and adding the prestige of a computer), and thereby conveying an 'apparently detached neutrality' which is 'as illusory as it is pervasive'. Like Malthus, the MIT study also underestimates the possibilities of continuous technical progress - that is, of obtaining greater outputs from the same inputs. Freeman quotes Golub's contention that the *Limits* is inherently dangerous because it gives a spurious appearance of precise knowledge of quantities and relations which are unknown (and unknowable); it stimulates gross oversimplification through aggregation and the simplicity of the computer models and mathematical techniques; and it makes it difficult for the non-numerate to rebut tendentious political assumptions.

There are those who would level very similar criticisms at Malthus himself, as well as at the neo-Malthusians. To evaluate their accuracy we shall have to go further, and critically examine the 'political assumptions' or ideologies and philosophies of ecocentrics, old and new (Chapters 5 and 7).

4.2 DARWIN, THE WEB OF LIFE AND ECOLOGY
4.2.1 Darwin as a Scientist
Along with Malthus, Charles Darwin can be regarded as a towering 19th-century figure whose ideas inform, among other things, the scientific strand of modern ecocentrism. Though the history of the ecosystems idea and general systems theory and philosophy is complex and shows many influences, one can recognise in Chapter 3 of *The Origin of Species* the fundamental outline of these powerful notions. The

outline of a systems view of nature was but one of several highly influential concepts which descended from Darwin. Those of competition, a struggle for existence in which the fittest survive, and, through this process, natural selection, constituted others. But although the concepts fit together in Darwin's work, where interdependence and interconnectedness may be a function of antagonistic relationships (e.g. predator/prey in the food chain), they might not fit together when they are developed and applied to society. Thus Capra (1982) in writing of the systems view of 'life' (including human society), takes the notion of interdependence and emphasises not its competitive but its co-operative implications. He says, 'The more one studies the living world, the more one comes to realise that the tendency to associate, establish links, live inside one another and cooperate is an essential characteristic of living organisms'. Once Capra has established this, he then rejects *Social Darwinism*, which emphasises an interpretation of nature as 'excessive aggression, competition and destructive behaviour' and chooses to project this on to human society.

As with Malthus, then, it soon becomes difficult for us to focus exclusively on Darwin's scientific theory and the supporting evidence, in any kind of 'objective' fashion. The ideological and philosophical *implications* of what they say soon become important factors in evaluating the work of these people, and its impact on man-environment attitudes. In examining it (see 5.3.3) we have to take into account the material position of the writers and their interpreters, and how that might affect their interpretations. However, we can start by attempting in this chapter a review of the ecosystems idea itself, as it developed from the work of Darwin.

This work would seem to have been in the true tradition of Bacon's man of science. It appeared in the wake of the advances, described earlier, of what Capra calls the 'Newtonian-Cartesian paradigm'. It also was founded on an age of discovery starting in the 16th century, in which both the boundaries of the known world were extended and the amount of knowledge of it was greatly increased. Much of this knowledge amounted to classification of animate and inanimate nature. Naturalists, zoologists, botanists, geologists - all were engaged in a great empiricist exercise of observing, recording and classifying natural phenomena. Such an exercise led to major works - for example by Buffon, Linnaeus and von Humboldt - and from them could come theoretical development. This process of deriving general principles - laws of nature - from empirical observation sounds like induction ; the method championed by Bacon. And Darwin seems to have been following it, basing his major works on observational evidence such as that which he himself gained from the voyage of the 'Beagle' and his work on coral reefs. Although Hull (1973) argues about whether Darwin was inductive or deductive, or both, Darwin himself certainly suggests the former when he says, of his *Origins*: 'on my return home, it occurred to me, in 1837, that something might be made out of the question by patiently accumulating and reflecting on all sorts of facts... After five years' work I allowed myself to speculate on the subject...' (Darwin

1885 p1). He published the *Origins* in 1859, and that was before he had intended.

4.2.2 The Web of Life

Some of Darwin's observations were set out in Chapter 3 of *Origin of Species*, entitled 'Complex relations of all animals and plants to each other in the struggle for existence'. He noted how, when sandy heath in Staffordshire was converted to coniferous wood, changes in bird and insect life followed. In Farnham, he remarked, the exclusion of cattle from heath by enclosure had allowed tree seedlings to generate where previously there had been only grass. 'Here we see', he said, 'that cattle absolutely determine the existence of the Scotch Fir...'. Furthermore, 'in several parts of the world insects determine the existence of cattle'. As an example, he traced linkages in what we might now call a Paraguayan 'ecosystem'. Describing parasitic flies which laid eggs in the navels of newborn cattle, horses and dogs, he wrote, 'The increase of these flies, numerous as they are, must be habitually checked by some other means, probably by other parasitic insects. Hence, if certain insectivorous birds were to decrease in Paraguay, the parasitic insects would probably increase, and this would lessen the number of the navel-frequenting flies - then cattle and horses would become feral [wild] and this would certainly greatly alter the vegetation: this again would largely affect the insects: and thus... the insectivorous birds, and so onwards in ever-increasing circles of complexity'.

Here are concepts of interrelation and interdependence, and Darwin goes on to present us with the idea of equilibrium, when he discusses the relationships arising from competition and the struggle for survival. In the long run the forces are 'so nicely balanced that the face of nature remains for a long time uniform, though assuredly the merest trifle would give the victory to one organic being over another'.

Intricacy of relationships is another dominant theme, and Darwin remarks that 'plants and animals, remote in the scale of nature, are bound together by a web of complex relations'. He illustrates this by describing the relationships between cats and flowers. Cats influence the number of field mice in a district. The mice influence the number of bees, because they destroy combs and nests, and bees, through pollination, control the frequency of heartsease and red clover. Thus:

> The dependency of an organic being on another, as of a parasite on its prey, lies generally between beings remote in the scale of nature...the structure of every organic being is related, in the most essential yet often hidden manner, to that of all the other organic beings with which it comes into competition for food or residence, or from which it has to escape, or on which it preys...

And we are greatly ignorant about the 'mutual relations of all organic beings...', partly because of the complexity of the relationships, which are harder to establish than are those governing non-organic objects:

> Throw up a handful of feathers, and all fall to the ground according to definite laws; but how simple is the problem where each shall fall compared

to that of action and reaction of the innumerable plants and animals which have determined, in the course of centuries, the proportional numbers and kinds of trees now growing on the old Indian ruins.

4.2.3 Human Ecology

Stoddart (1966) points out that Darwin's ideas, described above, were developed into Haeckel's new science of 'ecology' - a term used in 1869 - and that they culminated in Tansley's idea of the *ecosystem*. Darwin's particular contribution to the development of the idea of a 'web of life' was that he included *man* in it - the obvious implications of his evolutionary theory were that man had a common origin with the rest of nature. From about 1910, says Stoddart, the term 'human ecology' was used for the study of man and environment together - not in the sense of suggesting that man was determined by his environment, but in implying that he was not apart from nature; that he had a place in the web of life or the 'economy of nature'. In 1923, Barrows called on the Association of American Geographers to make geography: '...the science of human ecology... Geography will aim to make clear the relationships existing between natural environments and the distribution and activities of man... from the standpoint of man's adjustment to environment rather than environmental influence' (Barrows 1923). The organising concept of this paradigm for geography was to be the *region*, in which the geographer synthesised elements of the physical environment with human activity to produce the image of an individualistic and well-defined spatial entity with a unique character, a particular living ecosystem. Geography as human ecology thus saw the region and later the *state as a living organism* - another application of the organic analogy which, as we have noted earlier, runs through from medieval perceptions of nature to 18th-century neo-classical notions of senescence, to the 20th-century ecocentric revival of 'Gaia'. Stoddart says that the region and state as an organic ecosystem possessed 'properties of organisation of constituent components into a functionally related, mutually independent complex which in spite of continuous flows of energy and matter is in apparent equilibrium, and which possesses properties as a whole which are more than the sum of the parts'. Here, perhaps, is a reference to such intangibles as regional or national flavour and landscape character.

4.3 ECOSYSTEMS AND ECOCENTRIC PHILOSOPHY
4.3.1 The Meaning and Message of a Systems Approach

The characteristics of a systems approach to the natural environment are documented by Russwurm (1974). He says that the approach stresses that each part of the natural environment is related to each other part. The five subsystems - weather/climate; water; landforms; soils; biota - are parts of a larger ecosystem, whose overall characteristics amount to more than just the sum of the characteristics of the parts. This particular scientific way of conceiving of nature is therefore very different from that of Baconian science. Rather than seeking to break the 'machine' into its component parts, it seeks to study how the parts work *together*. Both living and non-living components of the ecosystem are unified by this monistic concept, which embodies a holistic approach to nature, reminiscent of the Chain of Being (see 3.1). The biotic and abiotic components are related and unified by virtue of the flows of energy and matter which pass through them.

Inputs of energy and matter are transformed by systems and subsystems into outputs of energy and matter. Transformation occurs via a series of processes or a succession of events that are characterised through the form of the energy which is used in them. Thus, the process of erosion which changes the form of rock in diagenesis depends upon the provision of potential energy from gravity and kinetic and heat energy from water, wind and ice. Energy and matter flow along pathways within a system before leaving it, and for an open system there is much exchange of matter between it and the environment, whereas a closed system is characterised by maximum recycling of material - that is, matter's transformation through various stages until it is available for re-use in its original form (viz. the nitrogen and other nutrient cycles). Systems growth occurs by an increase in inputs or more efficient use of energy and matter.

Unlike matter, energy cannot be re-used, because when it is used it changes to a different form (i.e. from potential to kinetic as a river flows downhill). Each change of form in matter also involves some dissipation of energy as heat. The laws of thermodynamics say that in a system of constant mass energy cannot be created or destroyed, but only transformed. Also, that as work is done energy is dissipated as heat, and to be available for work energy must be concentrated rather than randomly dispersed. Random dispersal, or disorganisation - known as entropy - is a state towards which nature tends. However, in an organised system there is efficient use of energy and matter and minimum entropy. Mature ecosystems (e.g. Appalachian forests) display high organisation (i.e. minimal entropy) because they are more diverse than immature ecosystems. They have more species and more niches are filled, and they are more able to capture matter and slow down energy dissipation. As Russwurm says 'Apparently, diversity and stability go hand in hand in natural systems', because diverse systems offer more resilience in the face of changing environmental influences. Thus a disease which affects one particular species (e.g. Dutch Elm) will wipe out a community consisting largely of that species. However, if it attacks a community consisting of many different tree species then the community can survive its losses. Together, then, stability and diversity imply organisation and some kind of equilibrium over time. Herein lies the root of ecocentrics' objections to so many aspects of contemporary economic organisation. It tends to lead to simple uniform ecosystems (e.g. monocultures in agriculture, standardised human communities) and the scientific laws of ecology decree that such uniformity is unstable in the long run. (However, as already noted, this sentiment is at variance with more romantic ideas which reject social complexity for perceived natural simplicity.)

Equilibrium, then, is needed for the continued existence of a system. This does not imply *unchanging* existence, for natural equilibrium is rarely static. Systems change and evolve while maintaining equilibrium; thus they display *dynamic equilibrium*. Typically one sees short-term fluctuations around a mean or equilibrium point which itself is moving. Two types of mechanism are involved here. If disruptions occur, the system tends to be brought back to its moving equilibrum point via 'negative feedback', that is, some kind of mechanism which is triggered by a change but whose effect is to nullify the change and bring the system back to its pre-change position. The Malthusian positive checks are examples. Increased income to the poor is supposed to lead to population increase, but this means that the income must be shared by more people. The size of individual shares goes down and the result is an increase in 'misery' (poverty) - which increases the death rate - bringing down, again, the population level. As discussed above (4.1.2), demographers have observed that this process does not actually happen. Instead, a 'demographic transition' occurs, in which additional income is not translated into more people. Population rates stabilise or fall with economic growth and the 'system' adjusts to a new equilibrium level of affluence. A permanent change has occurred and the mechanism that affects this is known as a 'positive feedback'. Long-term systems equilibria - the points about which short term fluctuations oscillate - also change therefore. They are pushed by positive feedback across 'thresholds' or critical limits. Once thresholds are crossed then stability occurs around a new equilibrium. The degradation of tropical forest to savannah, or savannah to semiarid steppe via overgrazing pressure and/or artificially-induced fire is an example of transition through thresholds towards new equilibria. The term 'degradation' implies increased entropy, greater simplicity and less resilience. It is a value-laden term, involving values in contradistinction to, for example, those of the Green Revolution evangelists. The science of ecology - apparently unlike the 'value-free' technocentric science which we have described - is replete with overt value-laden terms and with normative prescriptions for mankind. Thus, Russwurm, typically, says:

> Man's attempts to manage natural systems usually involve manipulation or buffering of inputs that either reinforce or work against tendencies existing in the system. Before undertaking such manipulation man should understand how a given input is assimilated or how it changes a system. Understanding feedback mechanisms is a prime requisite before we attempt to manage or change any facet of a natural environment.

Such understanding is based on careful conceptual, mathematical or empirical modelling of the real world. It demands 'a sound theoretical framework and solid empirical knowledge' (Russwurm 1974 p15). Similar admonitions are found in most basic ecology texts. The message is clear. It is that *for pragmatic reasons* (for fear of setting off adverse positive feedback) we must strive to know nature, using the rigorous procedures of scientific discipline and the conceptual framework, or paradigm, of ecology. We must recognise that whatever we do to nature will rebound on us, and, by extension, if we know about the effects of mistreating nature we will not mistreat it. *In contrast to the Baconian scientific creed, the message is that scientific knowledge should lead not to the exercise of power over nature, but should encourage us to seek harmony with it.*

This message is also very different from the romantic view that we must strive to know nature subjectively in order to attain spiritual gratification and ennoblement, and that we must respect nature *for its own sake* because it has a right to autonomous existence. However, the ecological scientific and non-scientific romantic messages about what should be our attitude to nature are not necessarily incompatible, even though the approaches to knowledge of nature may be. This is why the two can collectively form the *corpus* of ecocentric ideology. To the ecocentric, scientific and non-scientific views of the world may not be antagonistic - at least at first sight, and depending on whom the practitioners are. They *can* be regarded as mutually reinforcing, though there are those (e.g. Bookchin) who wish to appropriate some aspects of the ecosystems approach for environmentalism while rejecting others (we will deal with them later). By contrast, for the technocentric there does not seem much doubt that scientific and non-scientific views of nature are not usually compatible. The reductionist and 'objective' hallmarks of their science separate them from nature in such a way that simultaneous spiritual unification with it is rendered very difficult.

Stoddart (1965) has also documented the nature of the 'ecosystems approach', and in his view of its method it seems strongly to parallel the scientific method associated with the Newtonian-Cartesian paradigm. It also becomes clear that to Stoddart the conception of nature which is appropriate to the systems view still involves the machine metaphor.

Stoddart sees ecosystems as rationally and comprehensively structured. The structures, such as trophic levels and food chains, are therefore susceptible to rational investigation, and quantitative definition. Relationships between components of a system are defined in terms of energy and matter transfers. They involve interactions and exchanges which can theoretically be precisely quantified. Finally, systems obey laws, so that systems science can be nomothetic. The laws may be elucidated via model building, in which simple models can systematically be made more complex with increasing knowledge.

Stoddart sums up the ecosystems approach as follows:

> The study of the ecosystem, however, requires the explicit elucidation of the structure and functions of a community and its environment, *with the ultimate aim of the quantification of the links between the components...* partaking in general system theory, *the ecosystem is potentially capable of precise mathematical structuring...* Ecosystems are *ordered arrangements of matter in which energy inputs carry out work* (Stoddart 1965, emphases added).

One problem here, as Stoddart partly acknowledges, comes when *man* is incorporated into the system, as he must be when we build on the original Darwinian concept. If, as ecocentrics will have it, man is part of nature, then in scientific ecocentrism human beings become biological systems, and, as Stoddart says, human groups are 'highly complex systems...'.

If we view men and women as 'ordered arrangements of matter in which energy inputs carry out work', and the relationships between them and nature are seen as ultimately quantifiable, then difficulties immediately arise. Such a view is *not* compatible with the romantic view of nature and human beings. The new holistic, synthetic, ecologically harmonious science of ecology begins to become as suspect as the Newtonian-Cartesian paradigm to the romantically inclined ecocentric mind. Even though there is a message in it of harmony with nature, ecosystems harmony becomes a mechanistic, soulless harmony if it is interpreted as Stoddart interprets it. The ecocentric's scientific and non-scientific schizophrenia begins to look as if it might, after all, tear him apart.

To see if this rather serious prognostication need necessarily follow, we should now look more closely at objections to the ecosystems approach as a philosophy which can incorporate both man and nature in its monistic view. This examination will at the same time make us consider just how muuch a creature of free will in relation to nature man is. It may also suggest, however, that compromise is possible between a scientific ecosystems view of man and nature and humanistic and mystical approaches provided that the metaphor of nature is changed from that of a machine to an organism again.

4.3.2 For and Against the Mechanistic Ecosystems Approach

Stoddart thinks that to place knowledge about man and nature within an ecosystems framework would enable such knowledge to be precise, quantifiable, and amenable to expression and manipulation via the language of mathematics. This would have all the advantages associated with 'classical' science. The empirical study of 'facts' - e.g. primary qualities of size, shape, position, momentum - could lead to law-formulation, while hypotheses could be tested against reality via rigorous experimentation. Ultimately prediction and control could be arrived at. But at the same time the disadvantages of a reductionist approach are avoided. Such an approach could lead to man's abuse of nature, because of its insistence on separating the two. However, the ecosystems framework unites man and nature as part of the same 'machine'. Therefore man can be studied scientifically, as can the man-nature relationship. The question arises, as we have already suggested, as to how we then cope with aspects of human behaviour and characteristics which are not governed by rationality, and are not predictable - the irrational, emotional, spiritual, metaphysical and subjective side of man.

This is a familiar problem in geography - traditionally torn between the humanities and the sciences - and many, for example Moss (1979), believe that it can be dealt with in a scientific systems framework. Discussing the decision-making processes (which lead to man's appropriation and manipulation of nature in particular ways), Moss refers to the argument that human beings reach decisions not on the basis of what a situation 'in fact' is, but 'on the basis of what they think it to be'. This leads to the need for geographers to study and account for the effect of the 'perceptions which influence the decisions rather than the actual environmental stimuli which occasion them'. The particular way that groups and individuals perceive their environment involves their (shared) beliefs and values, ideologies and philosophies. These constitute an element in perception and decision making which, as we have said above, may not be governed by rationality. It will not be measurable or predictable. Moss proposes dealing with it by regarding it as an 'indeterminate element' within a set of elements that *can* be determined (analogous with the background 'noise' that is also received when an intelligible radio signal is received). Such indeterminate elements lend a degree of randomness or unpredictability to what can otherwise be predicted, and lead not to *certain* predictions but to predictions of the *probability* of some course of action being taken. The firmness of that probability is an inverse function of the random element. This randomness *is* amenable to manipulation in a mathematical framework. Stochastic (governed by the laws of probability) techniques have been evolved to deal with it. So:

> the problem of perception can be reduced to problems of indeterminacy and become more tractable in consequence. This may be achieved by specifying a culture framework within which the decisions are taken, and by further specifying a screen of values through which the environment is perceived: deviations from the norm can then be taken into account by allowing a degree of randomness in our methodological structure... This is more in accord with the traditional geographical approach and also more amenable to rigorous scientific treatment (Moss 1979).

The subjective side of man, then, is to be regarded as a random statistical element in an otherwise predictable model - a deviation from the norm. There is no difference between this approach to man and that of positivist social science (see 2.5).

Armed with this conceptual tool, we can approach the study of geographical regions. In a similar manner to Stoddart, Moss suggests that regions, and the people, economics and cultures they contain can be thought of in terms of 'energy use and exchange, the cycling and movement of materials, the character of functional niches and general organisation and so on... This would lead to a distinct science of regional systems within geography'.

This approach to societies is extended and amplified by Goldsmith (1978) who sees them as exactly analogous to biological organisms such as cells or other ecosystems - 'units of behaviour within the biosphere'. Cybernetics has shown that there is only one way to control a system, and that is via a *control mechanism* 'which operates by detecting data essential to the maintenance of the system's stable relationship with its environment...' The data can be resolved into a 'model or set of instructions' which ensures the normal day-to-day behaviour of a biological organism *such as a dog or man*' (emphasis added). For the social system such a model or template is the society's world-view, and the control mechanism for this, says Goldsmith, is *religion*. Religion can ensure that society's basic structure is maintained - religion 'admirably satisfies cybernetic requirements'. It encourages some order in 'traditional' societies, which unlike our own 'unstructured undifferentiated aggregation of people' display minimum entropy. Religion gives us the stability which the social ecosystem needs; it ensures that everyone can have a 'complete model of the environment and a corresponding strictly prescribed behaviour pattern'. In India, the caste system, which 'supplies a religious basis for inequality' has led to the maintenance of ...'a stable plural society in the face of overwhelming odds'. Religion consecrates or sanctifies the generalities of a society's behaviour pattern: 'if these generalities were to be disrupted, the consequences would be far-reaching and potentially catastrophic...'

Religion governs the behaviour of 'tribal man', who does not treat his environment simply as a resource to satisfy short-term needs. Through animism and the like he sanctifies it. But the absence in industrial society of a religion which provides a goal structure enabling it to achieve a stable relationship with its environment drives that society towards discontinuity and disintegration (entropy). Goldsmith forecasts that with increased industrialisation and increased recognition that a materialist paradise is unattainable will come 'a growing number of Messianic movements which will attempt to establish a new social order based on a new view of man's relationship with his environment. Many of them will adopt at least a facade of Christianity...' In systems terms, religion will be reestablished as a control mechanism enabling the social system to reach equilibrium and stability with respect to the environment.

There are several objections to these 'systems' views of man, society and nature. Some see them as dehumanising and grossly mechanistic, while others emphasise their essential determinism, placing man in too subordinate and dependent a role in relation to nature.

Can relationships between men, and within and between societies, and between man and nature, be precisely quantified and mathematically structured? Can culture and values, emotion and irrationality be built into a predictive social model as mere 'degrees of randomness'? Can religion and spirituality be regarded simply as a control *mechanism* holding a social system onto a particular course? Are men and society qualitatively no different from cells or dogs? Bookchin disagrees. He believes that a systems view of man and society is no better than that of the Newtonian-Cartesian paradigm, in which the analogy of nature as a clock is extended - through social sciences - to people. In such a mechanistic view other people become mere objects, separated from the subject, and predictable through rational laws:

> If energy becomes a device for interpreting reality...we will then have succumbed to a mechanism that is no less inadequate than Newton's image of the world as a clock...Both reduce quality to quantity...both tend towards a shallow scientism that regards mere motion as development, changes as growth... systems analysis reduces the ecosystem to an analytic category for dealing with energy flows as though life forms were mere reservoirs and conduits for calories, not variegated organisms that exist as ends in

themselves, and in vital developmental relationships with each other (Bookchin 1980 p88).

Bookchin (p278) illustrates this by quoting Buckminster Fuller's view of man as a 'self balancing, 28-jointed adapter-base biped, an electromechanical reduction plant, integral with the segregated stowages of special energy extracts in storage batteries' and of the nervous system as a 'universally distributed telephone system needing no service for 70 years if well managed' etc.

Von Bertalanffy (1968), an advocate of general systems theory, is nonetheless also aware of these criticisms of its mechanistic potentialities, and he sums them up no less eloquently:

> The dangers of this new development, alas, are obvious and have often been stated. The new cybernetic world, according to the psychotherapist Ruesch...is not concerned with people but with 'systems'; man becomes replaceable and expendable. To the new utopians of systems engineering,... it is the 'human element' which is precisely the unreliable component of their creations. It either has to be eliminated altogether and replaced by the hardware of computers, self-regulating machinery and the like, or it has to be conformist, controlled and standardized. In somewhat harsher terms, man in the Big System is to be - and to a large extent has become - a moron, button pusher or learned idiot, that is highly trained in some narrow specialisation but otherwise a mere part of the machine. This conforms to a well-known systems principle, that of progressive mechanization - the individual becoming ever more a cogwheel dominated by a few privileged leaders, mediocrities and mystifiers who pursue their private interests under a smokescreen of ideologies (Von Bertalanffy 1968).

In Chorley's (1973) view, the (mechanistic) ecosystems approach's inadequacies lie in the answer which it gives to the philosophical dilemma of the nature of the relationship between man and nature. He says 'The ecological model may fail as a supposed key to the general understanding of the relations between modern society and nature, and therefore as a basis for contemporary geographical studies, because it casts man in too subordinate and ineffectual a role'. Chorley maintains that whether we like it or not, the realistic picture of the man-nature relationship is one in which the former dominates and moulds the latter to his will. In the manner of a spiteful slave, nature might bite back at him in exchange for his abuses, but man is ultimately the master. Or, to put it in systems terms, he is the control in a control system. He impels the system, via the positive feedbacks he induces, through thresholds into new equilibria with the environment. This realistic view of man's increasing hold over nature in an industrialised society is belied by a conventional systems approach.

4.3.3 Environmental (Geographical) Determinism

What Chorley is contradicting here is the implicit idea in the mechanistic systems philosophy, and in ecocentrism as a whole, which is that because man is *part* of a system there are *limits* to his degree of control over what he can do in developing his society through the appropriation of nature. As Learmonth and Simmons (1977) say, 'there is an overall ecological "envelope" which sets limits to human endeavour. Because, in the systems view, our exploitation of nature will rebound on us to our detriment, then the course of human development is circumscribed. At each trophic level in an ecosystem limited energy is available to support life. Space will be limited; so will the rate of nutrient recycling. All these ideas are encompassed by the biological phrase for limits, that is, 'carrying capacity'. Learmonth and Simmons say: 'There is no need to labour... the parallel with the human situation. For a long time we have escaped limits because we could always enlarge the carrying capacity... the stretching process is about to stop'. This is because the industrial way of life is contaminating ecosystems and diminishing their carrying capacity. The general Malthusian message of *Limits to Growth* holds 'rather like applied common sense: there is a limit to the carrying capacity of the world for human beings...'

This is the modern ecocentric version, put in scientific terms, of the old idea (going back to Hippocrates in essence, see Glacken 1967) that the environment itself is 'capable of determining the course of human development' (Pryce 1977). Ritter was one of many geographers who preached it, in the form of *environmental determinism*.

In its crudest form it often led to facile and immoderate generalisations, which are well illustrated from the writings of Kant, Reuter, Buckle, Ratzel, Demolins and the like by Tatham (1951). The following well-known example of Ellen Semple's work illustrates the mechanistic cause-effect relationships of crude determinism:

> Man is a product of the earth's surface. This means not merely that he is a child of the earth, dust of her dust; but that the earth has mothered him, fed him, set him tasks, directed his thoughts, confronted him with difficulties that have strengthened his body and sharpened his wits, given him his problems of navigation, of irrigation, and at the same time whispered hints for their solution. She has entered into his bone and tissue, into his mind and soul. On the mountains she has given him leg muscles of iron to climb the slope; along the coast she has left these weak and flabby, but given him instead vigorous development of chest and arm to handle his paddle or oar. In the river valley she attaches him to the fertile soil, circumscribes his ideas and ambitions by a dull round of calm, exacting duties, narrows his outlook to the cramped horizons of his farm. Up on the windswept plateaux, in the boundless stretch of the grasslands and the waterless tracts of the desert,...where the watching of grazing herds gives him leisure for contemplation, and the wide-ranging life a big horizon, his ideas take on a certain gigantic simplicity; religion become monotheism, God becomes one, unrivalled like the sand of the desert and the grass of the steppe,... (Semple 1911 pp1-2)

Tatham points out in defence of these and other extreme standpoints that Semple and other early determinists were working with limited data, and that they tended to overstate their case in order to prove a theory and give prominence to factors of the environment which had hitherto been neglected. James (1972) however, counters that the arguments used were so abstracted that they were incapable of objective verification.

By contrast, the 'scientific' determinists of the early 20th century - of whom Ellsworth Huntington and Griffith Taylor are foremost examples - based their work on empirically-derived data. Huntington thought that human history could be explained and interpreted substantially in terms of world climates, which, with biological and cultural inheritance, determined the nature of human societies and the degree to which they were 'civilised'. Thus, he looked at maps showing features attributable to climatic change and concluded that desiccation had occurred in Central Asia during the time when military expansion had taken place from this region into Europe (Huntington 1907). From this alleged correlative relationship he went on to assume a *causal* one - which, of course, is not regarded today as sound scientific practice. His most famous books, *Civilisation and Climate* (1915) and *Mainsprings of Civilisation* (1945) carried on with the correlations. On the basis of data on work performance in factories in the USA he concluded that there was an 'ideal' climate most conducive to high levels of physical and mental activity. Unsurprisingly, this climate was that which was characteristic of the temperate regions of the world, including Huntington's own cultural region of Anglo-America (the passage of alternating depressions and anticyclones over these areas imparts climatic variety which was held to be a further stimulant to physical and mental activity). Northwest Europe, eastern USA, southern Canada and BC and Japan were the areas of most stimulating climate, where it was to be expected that people would be most energetic. Again unsurprisingly, these and other 'favourable' regions matched up with those which were regarded as most 'civilised' (obtained by questioning 'experts' such as (western) anthropologists and historians on where it was that they discerned most social, economic and political development). In these exercises, Huntington was doing what, according to Glacken, has been done by all climatic determinists since the Greeks. He was showing that his *own* region was somehow the one most conducive to the development of a superior culture.

Griffith Taylor's approach was different, and more cautious. He was concerned with evaluating into what areas his own culture might expand. He mapped climatic and resource data so as to be able to identify regions suitable for white settlement (Taylor 1937). His studies were, then, predictive, and in postulating where white settlers could go in Australia (Taylor 1940) and how many that country could accommodate he was more conservative than the Australian authorities who wanted to attract white immigrants. His basic thesis was that the *direction* of man's development was subject to overall environmental constraints, but that the *rate* at which such development took place was subject to human control. Humans were like traffic controllers, who could not alter where traffic went, but could control its speed. Thus, this 'stop-and-go determinism', like the climatic determinism of Huntington, cast man in a less 'subordinate and ineffective' role than did Semple.

A greater philosophical departure from crude determinism was present in 'possibilism', where it was held that although the environment set overall constraints to human endeavour it offered a range of *choices* to man on his courses of action. Tatham points out that this view attributed more importance to man and less to environmental influences. Thus Febvre, who coined the term 'possibilism' said 'There are no necessities but everywhere possibilities; and man as the master of these possibilities is the judge of their use... It has been the custom for many years to speak of human society in the great climatico-biological regions as adjuncts, so to speak, of plant and animal societies which were themselves, it was assumed, strictly dependent on meteorological phenomena. But these regions have nothing tyrannical or determinant about them'... (Febvre 1924, quoted in Tatham 1951). As Tatham says, such quotations make it quite clear that for possibilists 'nature does not drive man along one particular road... it offers a number of opportunities from which man is free to select' (Tatham p155). However, within this analysis there still lies a fairly limited view of the power of man - he is choosing between possibilities offered to him *by* the environment. While there is more emphasis on man, it is still within an overall deterministic framework.

We have asserted that the philosophy of environmental determinism underlay the Malthusian view of the man-nature relationship and has been implied in the development of the mechanistic 'ecosystems' approach. Thus, two major influences upon current ecocentric thinking can be regarded as essentially deterministic. These influences are clearly discernible in the writings of neo-Malthusians Hardin and Erhlich, in major works like *Limits to Growth* and the *Blueprint*, and in the views of ecocentrics like Goldsmith (1975B), who still, after Huntington has been discredited for many years, attributes the ebb and flow of civilisation to climatic changes. (Environmental determinism still underlies some modern geographical work as well, see Maunder 1970 and Norwine 1981.) However, there are other influences on ecocentrism which are less deterministic. One of these is the 'organic' ecosystems approach, which we will describe briefly below.

4.3.4 An Alternative ('Organic') Systems View
The criticisms of the 'ecosystems' approach which are described above relate to a particular *kind* of systems view - that is, as we say, a mechanistic one. It is a conventional systems view, but there are those who regard it as an inaccurate representation of systems reality. In their opinion a more appropriate analogy for the system and for its components, including nature, would be that of an *organism*. This is because the organic analogy would be more able to admit of and account for the irrational, spiritual, metaphysical aspects of man and his relationship with nature. In it the way that we relate and respond to nature, and shape it, is more than just mechanical, measurable and deterministic in the senses described here. And how we relate to each other cannot be described in the more utilitarian terms employed by Goldsmith. Man and nature are one and indivisible - there is no separation between them as there is in the Cartesian dualism of classical

science. A similar subjective relationship between people is also emphasised, and an organic systems view is heavy with subjective messages about how we should live. Above all, it opposes the reductionism of classical science, whereby living and non-living things alike can be reduced to matters of chemistry, then physics, then mathematics. Instead it asserts the doctrine of *vitalism*, that there is a vital principal to living things, that cannot be reduced to physics and chemistry and separates them fundamentally from non-living things.

Von Bertalanffy (1968) clarifies this perspective. General systems theory contains a synthetic, holistic view of man and nature, in which *relationships* are emphasised and studied. It is a scientific exploration of the relationships and of '"wholes and wholeness", which not so long ago were considered to be metaphysical notions transcending the boundaries of science'. Whereas classical science is analytical, mechanistic and one-way causal (deterministic), the organic systems view renders obsolete the physicalism, atomism and reductionist view of nature (to elementary particles governed by conventional laws of physics) of this 'objective' positivist science.

'Physics itself', says von Bertalanffy, 'tells us that there are no ultimate entities like corpuscles or waves existing independently of the observer' (see 5.2.2), and the organic systems view is also against theories which declare that reality is nothing but 'a heap of physical particles, genes, reflexes, drives or whatever...' It thus bridges the two cultures of humanities and science. The humanistic concerns of general systems theory makes it different from those of mechanistically oriented systems theorists.

There is a close parallel here with Capra (1982). He also rejects classical mechanistic, reductionist science, believing, too, that physics now shows us that there are no ultimate particles and there cannot be any fundamental distinction between the human and the physical world. Humans are part of larger self-renewing systems which amount to an organic whole - the planetary ecosystem or Gaia (Lovelock 1979). Systems are integrated wholes whose properties cannot be reduced - systematic properties are destroyed by dissection. When we look at man and nature within this framework we are making observations within the context of science but going far beyond science. 'Like many other aspects of the new (systems) paradigm', says Capra, 'they relfect a profound ecological awareness that is ultimately spiritual'. The new systems biology and the new physics have a growing affinity with mysticism, because the latter, in its Eastern form, enphasises the mutual interdependence of all of reality and it emphasises human consciousness and spirituality. In the new systems view organisms evolve to a degree where they transcend themselves. That is, they become capable of *organising themselves* and nature through the development of consciousness. Consciousness is a property of the higher animals, and the human mind is a 'multi-levelled and integrated pattern of processes that represent the dynamics of human self-organisation'.

This is a more sophisticated view of human beings than the mechanistic one will allow, and it is far less deterministic. '*Self-*

transcendence' and *self*-organisation through evolving consciousness are concepts which stand in contradistinction to determinism, where an organisation is *externally* imposed. They are concepts which have much in common with free will philosophies, and they lie, therefore, at the opposite end of a philosophical spectrum to determinism.

We now intend to examine that philosophical spectrum - which underlies ecocentric and technocentric ideologies - and to relate it to the concept of objectivity, in order to resolve a central paradox which has arisen.

CHAPTER 5
SCIENCE AND OBJECTIVITY

5.1 A PARADOX

We have seen that both ecocentrics and technocentrics make heavy use of science as a method for investigating nature and its characteristics. It appears that each side makes its investigations according to the canons of scientific method, and *then* draws conclusions about the nature of the man-environment relationship as it is and as it should be. Such conclusions are based, we assume, upon a rigorously objective scrutiny of the data, and a rational induction of principles or 'laws' from it. Through this process - which is compatible with Bacon's concept of scientific method and the popular modern view of how science proceeds - we might expect to come to some kind of agreed, objective and authoritative conclusions about man's relationship to nature as it is, can and should be. Impartial policies based on these conclusions could then be formulated in the interests of all, for social progress in general.

However, this is not what happens. Paradoxically, it seems, *both sides use science and scientific method to investigate nature but they come to very different conclusions about it* and what man's relationship to the environment can and should be. Technocentrics proclaim, in the manner of the Baconian creed, that scientific knowledge tells us that we can *manage, dominate and manipulate* nature for our own ends. Ecocentrics, from their systems perspectives, maintain that domination and exploitation must stop and be replaced by a relationship of *harmony and stewardship*.

Clearly, both sides cannot be right, and they cannot have reached their conclusions purely after objectively observing a monolithic world of 'plain, hard facts' and laws stemming from them - otherwise both sides would have been likely to reach similar conclusions. Indeed, these facts and laws would have meant that there would not have *been* two sides. Scientific knowledge would have produced a consensus: a unity. The lack of such unity suggests that in some way the process cannot have been purely 'objective'.

Perhaps each side, influenced by *preconceived* notions about what the man-nature relationship should be, decided to select different sets of 'facts' to sustain and strengthen their preconceptions. They might have been biassed to start with, and have used a veneer of scientific objectivity to legitimate - to make respectable and acceptable - an ideology which served the interests of particular groups in society, including themselves. We shall shortly examine how this might have been done over environmental issues, and this will lead us towards a Marxist perspective on these issues. But first we should consider another, perhaps more fundamental, proposition.

This says that in any case an objective examination of nature is actually impossible to achieve because there *is* no objective nature. That is, that the Cartesian idea of a world external to ourselves is fundamentally incorrect, and the subject (ourselves)/object (a physical world of fundamental measurable particles) dualism is not valid. If this were to be so, then of course a detached - objective - view of 'the environment', uninfluenced by the investigator's preconceptions, would not be achievable. To examine this proposition we must reconsider the fundamental philosophical issue which has already arisen, namely that of whether man's relationship to nature is a determined one, or whether it is a function of human consciousness and free will.

5.2.1 Determinism in Classical Science
The reason why we have to re-examine the question of determinism when considering the subject-object dualism is that this dualism, and the particular scientific view of which it is part, constitute a fundamentally deterministic view, at least in some senses.

We have described determinism, in section 4.3.3, in the context of the man-nature relationship. In it, man's physiognomy, character, psychology, economic, social and cultural activities, and ability to multiply can all be shaped or limited by nature. Man's freedom to act and develop is *circumscribed*. And his importance in relation to nature and natural laws is *diminished*. However, this is but one application of determinism; an extension of a more basic concept. For we are really describing a relationship (between nature and man) which is essentially one of *cause* and *effect* between two variables. In it, a change in the independent variable (nature) would produce a corresponding and predictable effect on the dependent variable (man), according to laws to which man and nature are subject. (One could think of the effects on human activity which would result from, say, climatic desiccation.) Being scientific laws, they operate uniformly in space and time and they cannot be denied. One could move away from classical environmental determinism and argue that, given sufficient knowledge of the laws, and ingenuity, man can manipulate them and put them to *his* use in exploiting nature, but despite this relative freedom of action, man still cannot act so as to suppose that the laws do not operate - it is beyond his control to change them: to change the basic 'facts of life'. There are other kinds of determinism, which say that men are subject to economic laws or laws of history, or to God's laws, but in them all the law-governed cause-effect relationship between variables is implicit. Essential to the relationship is the idea that there is a certain *separation* between the variables (man and God, events, nature). Of course they are related, but they are distinct from one another. Here is man, and there is nature, out there; separate from us and capable of being an independent objectively-known variable. There is, therefore, a subject-object distinction. Nature can operate and change independently and objectively - that is, separately from the intentions of human beings. Classical science holds that nature is a machine, behaving in a predictable way according to laws which determine it.

So this view of nature as a set of mechanical objects external to man conveys the notion of cause-effect relationships. God might be the ultimate cause, and work through nature's laws to have particular effects on man. In this way, as Malthus suggested, God would be the primary cause of the fate of human beings - he would operate via the secondary cause, of nature and natural law, to control human destiny (Glacken 1967). (This kind of explanation is a mix of determinism with teleology.) If human actions are thus included in the deterministic system it also follows by extension - 'though Hume and others have disputed this - that no one is morally responsible for his actions' (Bullock and Stallybrass 1977).

5.2.2 The Determinist-Free Will Philosophical Spectrum (see Figure 6)
(Parts of this section are based on material supplied and interpreted by Martyn Youngs.)
This contention is strongly disputed by adherents of *existentialism*, a philosophy which emphasises man's freedom to act independently of any laws, natural or otherwise, and according to his own choice. It is a *free will* philosophy which can be seen as 'a protest against views of the world and policies... in which humans are regarded as the helpless playthings of historical forces, or as wholly determined by the regular operation of natural processes' (Encyclopaedia Britannica 1978). This anti-determinist perspective is, again, connected with the question of objectivity. Existentialists hold that the only objective external fact independent of the control of human beings is the fact of their being brought into existence, and that death will one day come to terminate that existence. In all other respects it is *they* who are in control. There are no outside independent laws, of economics, history, nature or whatever, which they cannot deny or shape for themselves. Thus we are all free to choose how we will behave and develop and how we will shape society and nature. It follows that the consequences of our actions are down to *us*, and therefore if they are unpleasant that is our fault - not the fault of outside forces or external laws. There are no excuses for our not thinking and acting for ourselves and we are 'condemned' to be free because everything is theoretically permitted to us and we therefore carry a heavy burden of responsibility for what we do. As Sartre put it, 'Man is responsible for what he is. The first effect of existentialism is that it puts every man in possession of himself as he is, and places the entire responsibility for existence squarely on his shoulders' (Sartre 1943, see also Sartre 1946). We have a duty to face up to this, and not to base our actions thoughtlessly on what is said to be prescribed *for* us

('laws' of society, of economics, of nature). That is abdicating responsibility.

Clearly, this philosophy is very relevant to radical ecocentrics, who preach the need for value changes and consequent social change. It tells them that such change *is possible*: that we can reform ourselves and our society and thereby fashion a society and nature in such a way as to create a more harmonious relationship between the two. By contrast, conservative ecocentrics, who may emphasise a mechanistic systems view, will argue more deterministically, that it is we alone who must change in response to natural law. We must accept *limits* to our freedom of action imposed by nature. And most technocentrics will go further than this, maintaining that no fundamental changes are desirable or even possible on either side.

The interplay between these ideas and the concept of objectivity can be seen more clearly by reference to the free-will philosophy to which existentialism is related - that is *phenomenology*. It is summed up by Nietzsche, who said, 'Objectivity is the main enemy of understanding. It means the myth that there are hard observable facts...but all the concepts we employ in describing the world and predicting its behaviour are imposed on it by ourselves. We have a choice about what view of the world we adopt' (quoted by Warnock 1979 p13). Phenomenology is not just a philosophy which implies that we are not subject to external laws, imposed by and through forces independent of ourselves. It holds that there actually *is no world external to and separate from ourselves*. There is no 'reality' in the sense of there being an external 'real world' divorced from our own consciousness and capable of existing if we did not exist. Nature cannot exist without ourselves also existing - if we ceased to be, so would it cease, and vice-versa. Therefore what 'the environment' is, and is like, is a function of our own subjective construction of it. The very use of such terms as 'ourselves' and 'human consciousness' in opposition to terms like 'nature' and 'the environment' is inappropriate and shows the extent to which we are immersed in positivist science as our way of seeing the world. For the phenomenologists these terms are meaningless because there is *no separation* between ourselves and a 'nature' or a 'reality'. We and the world are one - a single united entity. Phenomenology is thus anti-positivist science, opposing the Cartesian dualism. It does not deal in laws, or in cause-effect relationships consequent on the dualism, and neither can it be concerned with analysis - breaking the world into parts. It is taken up with the individual and the unique, and it is holistic in approach (see Walmsley 1974). Furthermore, 'Phenomenology is a philosophy of science based on the same foundations as idealism: all knowledge is subjective' (Johnston 1983).

It follows from this that if we want to know and understand nature we must, as Hume said in his *Treatise on Human Nature* in 1737, come to a full comprehension of *man*. For if there *is no* nature except as structured through human consciousness, knowledge and truth cannot exist independently of man. This, of course, is the ultimate expression of the free will idea. It says that we are not determined by nature - on the contrary we are so free in relation to nature that we can actually say that what nature *is* depends on what we will it to be - on how we want to structure it through our consciousness. Thus knowledge of nature has to be gained via an understanding of man's experience of the world, and only through his intentions and attitudes towards it (Tuan 1971A). This is the notion of *intentionality* which was formulated by Husserl. It holds that human intentions, working through consciousness, make the structure and shape of the world what it is for us - whether it is the world of 'solid objects' like buildings (Bach tells us, in his novel *Illusions* (1977), that, given faith, we can walk through them), or of metaphysical thoughts, ideas and values. Our consciousness takes in 'information' and makes it into 'the world'. In so doing it structures the world into *essences* of things. For example, if we take one part of nature such as a cat, we can see that every cat is different in some way from every other cat - yet it also has something in common with all other cats, which we can call 'cat-ness' or 'essence of cat'. This essence is the combination of characteristics by which we recognise cats and separate them off from other parts of nature, and ourselves. But the cat essence is an idea - *our* idea - which is different from our idea of, say, dog. Because it exists in human consciousness and is therefore not really separate from it it follows that we could reshape that idea and restructure the essence of cat into some other idea. In the same way we also create our own essence. Such creativity is clearly a function of our free will.

We are dealing with an old philosophical problem here (Russell 1946 p177), and not one restricted to phenomenology. But it gets us to that phenomenological concern which says that the process of learning about the world (nature) is a process of trying to understand how our consciousness imparts structures on to it (note that this very phraseology implies a division between man and nature which the phenomenologist denies). There are two aspects of this question.

One is that we may be concerned with processes that do not happen in any kind of thought-out way. We are not, in other words, 'conscious' of the fact that we are imposing structures. So the objects of phenomenological study may be those aspects of the world - or our surroundings - that we do not think about in everyday life but which are nonetheless extremely important to us. We are talking here about our 'lifeworld' (Seamon 1979A, Buttimer 1976) - the world of familiar ideas, experiences and objects, like the furniture in our rooms and the shapes, shadows and cracks on the walls and ceilings - on to which we do not consciously bring to bear our thought processes but in whose sudden absence we would feel disturbed, as if something were 'wrong' (Seamon 1979B).

The method of study of this lifeworld must be, therefore, to try to think consciously and descriptively about things we do not usually think of in this way, in order to bring them from the back to the forefront of cognition, and make explicit what was implicit. And since 'the lifeworld' is a personal thing, varying from individual to individual, we cannot induce law-like statements about it.

The second aspect of imparting structures 'on to reality' follows from this point. If we want to understand how individuals or groups (acting to a degree under a cultural consensus) impart *their* structures, then we must get rid of *our* presuppositions, which will colour our study of them. In other words, we must try to understand the world as they understand it and empathise with them. Our consciousness must become their consciousness - we cannot study them objectively because objectivity means detachment. Thus, if we want to know why individuals and groups have shaped their environments - e.g. their field patterns, gardens, cities - as they have, and why they have imposed the structures they have, we must experience their environments as they experience them. In this way we can translate more accurately these environments into the ideas which they symbolise (Tuan 1971A). This seems a practical and sensible thing to do, and it belies the notion that the science of phenomenology, unlike positivist science, is full of abstruse concepts that have no relevance to our everyday lives. Rather, it is the peculiar view of positivist science, which divorces man from the world around him and objectifies both, that may produce inherently irrelevant solutions to the problems of the man-nature relationship. To give one example of the use of phenomenological method in planning environments, Rowles' (1978) study of American senior citizens attempted to empathise with their lifeworld. It discovered that in it, for example, the preferred routes of travel between two points might well be longer and more 'roundabout' ones than those assumed by positivist planning science, because the routes would keep the travellers in sunlight, or give them somewhere to sit and rest on the way, or for less tangible reasons of sentiment or harking back to former experiences. Such a conclusion, which might not have come to light in an 'objective' study, has, of course, great practical implications for planning. It says that people's perceptions of places are partly a function of memories and associations derived from earlier experiences, and other 'irrational' considerations, and to discover what is important about places we must identify subjectively with these perceptions.

We have used the term 'science' of phenomenology, for science is what it is a philosophy of, even though its approach and methods are opposed to what we more usually regard as science. However, phenomenologists would say that classical, positivist, 'science' has appropriated this word, which means 'systematic and formulated knowledge', or just 'knowledge' (Concise OED). There is yet another 'kind' of science which also runs counter to the classical notions (though it was born of classical science), and tends to support implicitly the views of phenomenologists, organic systems advocates and mystics, (see Davies 1983) including Transcendentalists. This is the science of subatomic physics, whose language is really mathematics. It is therefore most difficult to understand its concepts via the written word, but the implications of the concepts have been admirably spelled out by Capra.

He describes the inadequacies of the Newtonian-Cartesian view of nature, that sees it as composed of objects external to ourselves which are made up of fundamental solid particles (atoms) that have objectively measurable primary qualities of size, shape, position and motion. Subatomic physics has shown us that, instead, the 'atom' consists of vast regions of space in which extremely small particles move. And the development of quantum theory showed that even these were not like the solid objects of classical physics:

> The subatomic units of matter are very abstract entities which have a dual aspect. Depending on how we look at them they appear sometimes as particles, sometimes as waves; and this dual nature is also exhibited by light which can take the form of electromagnetic waves or of particles... It seems impossible to accept that something can be at the same time a particle - i.e. an entity confined to a very small volume - and a wave, which is spread out over a large region of space... The apparent contradiction between the particle and the wave was solved in a completely unexpected way which calls in question the very foundation of the mechanistic world view - the concept of the reality of matter. At the sub-atomic level matter does not exist with certainty at definite places, but rather shows 'tendencies to exist' and atomic events... show 'tendencies to occur'.

These tendencies can be resolved mathematically as probabilities, hence;

> Quantum theory has thus demolished the classical concepts of solid objects and of strictly deterministic laws of nature. At the subatomic level the solid material objects of classical physics dissolve into wave-like patterns of probabilities... [Furthermore] All particles can be transmuted into other particles; they can be created from energy and can vanish into energy... classical concepts like 'elementary particle', 'material substance' or 'isolated object', have lost their meaning; the whole universe appears as a dynamic web of inseparable energy patterns... *The crucial feature of atomic physics is that the human observer is not only necessary to observe the properties of an object, but is necessary even to define these properties. In atomic physics we cannot talk about the properties of an object as such. They are only meaningful in the context of the object's interaction with the observer*(Capra 1975 pp69-72 and 144).

The emphases here are added, because this statement is very close to the ideas of the phenomenologists. Capra goes on to illustrate it by reference to observing a subatomic particle. One can choose to measure the particle's position and its momentum (mass times velocity). But Heisenberg's uncertainty principle says that:

> ...these two quantities can never be measured simultaneously with precision. We can either obtain a precise knowledge about the particle's position and remain completely ignorant about its momentum (and thus its velocity) or vice versa... The important point now is that this limitation has nothing to do with the imperfection of our measuring techniques... *If we decide to measure the particle's position precisely, the particle simply does not have a well defined momentum*... [And vice versa, such that the scientist] ...influences the properties of the observed objects... *The quantum principle... destroys the concept of the world as 'sitting out there' with the observer safely separated from it by a...slab of plateglass.* (Capra 1975 pp144-5, partly quoting Wheeler 1973).

So the concept of objectivity is again refuted, and the scientist becomes a participator, influencing the quality of what he perceives, not a detached observer. Richards (1983 p76) believes, too, that the fundamental discoveries of physics have blurred 'the traditional Western dichotomy between the objective and the subjective... If, in studying nature, science inevitably affects the nature of what it studies, it must be that science after all can give us only the appearance of the world and not its reality'. He contrasts Galileo's statement that 'The conclusions of natural science are true and necessary, and the judgement of man has nothing to do with them' with Heisenberg's conclusion that 'We can no longer consider "in themselves" those building stones of matter which we originally held to be the last objective reality... basically it is always our knowledge of the particles alone which we can make the object of science... Even in science the object of research is no longer nature in itself, but man's investigation of nature'. The latter is very close to the phenomenologist's position that there is no meaning in the notion of nature except as structured by man's consciousness.

Such a view might, as Tuan (1972) says, make man's abuse of nature 'easily understandable'. Because if there is no separate, independent nature in its own right then what it is and how it changes becomes a function of man's free will. On the other hand, these concepts are also compatible with the idea, in 'organic systems' and mysticism, of one-ness with nature - which leads to the opposite of an abusive attitude. This suggests that therefore the underlying philosophies themselves are not, in their interpretation, free from the presuppositions of their users. They emphasise certain aspects of the man-nature relationship, but *how* these aspects are interpreted into a specific view about what the man-nature relationship ought to be seems partly a function of pre-existing bias - a bias which stems perhaps from ideology. If this is so, then the philosophy of science can be appropriated (just, as we will go on to suggest, as its methods and results may be appropriated) by groups with different ideologies.

Different philosophies which are relevant to our views about the man-nature relationship are represented in Figure 6, which also shows intermediate positions along a thought spectrum with determinism at one end and free will at the other. As we move away from determinism, the importance of man in relation to nature increases, to such an extent that at the other extreme there *is* no nature distinct from what man wills it to be. So the subject-object dualism also collapses, and man and nature become one, not separate. The idea that there are scientific laws governing either, which is implicit in nomothetic positivistic science, gives way to ideas which are amenable only to idiographic (and predominantly descriptive) methods, and the positivist's rejection of knowledge gained through emotion and intuition is also overturned as we move away from determinism. Thus the free will end of the spectrum sets much store by aspects of humanity which are also found in mysticism (including romanticism) - and, indeed, because it places such emphasis on man and human intentions it has an affinity with humanism and the humanities which is lacking at the determinist end.

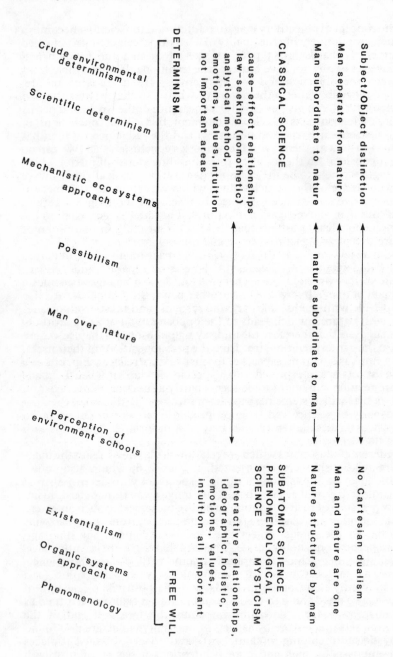

FIGURE 6   The Determinist - Free Will Philosophical Spectrum

When we examine some intermediate positions we can clearly see that although we might debate where precisely possibilism or the mechanistic ecosystems view ought to be placed, they are inherently more deterministic than not. And we would also argue, as suggested, that technocentric views of the man-nature relationship which see the latter as dominated and shaped by the former nonetheless have much in them which is deterministic. For although they of course emphasise man's freedom of will and action *over* nature, nonetheless they posit an essential *separateness* of the two - a dualism in which man is the 'independent' and nature the dependent variable and the former determines the latter (see Chorley 1973). This is really the position derived from the Baconian creed, arrived at through classical and fundamentally determinist science.

Another intermediate position characterised by a mixture of elements of both extremes is that which emphasises the importance of *human perception* in relation to nature. It is familiar as underpinning the study of natural hazard perception, for example, since World War Two, (White 1965 and 1973, Burton, Kates and White 1978). It says that decisions about whether to take action to avert a hazard are a function not of how potentially dangerous that hazard *is*, measured in any kind of absolute or 'objective' terms, but of how dangerous people *perceive* it to be - indeed, of whether they perceive that the 'hazard' is there at all. Thus the hazard of a Thames flood which would destroy large parts of London existed in 'reality' for hundreds of years before any action to build a flood barrier was actually undertaken - because the perception of the threat was not sufficiently great to impinge on political consciousness. We are really talking about *two* kinds of environment here. There is a 'real' environment of 'hard' physical 'facts' which is external to and independent of humans, and there is the environment 'as perceived' by humans. Ever since Kirk (1963) argued that the latter was more important than the former in shaping human decisions in space (and therefore in shaping landscape) behavioural geographers have worked to understand the process of human perception. Borrowing from behavioural psychology, which is a primarily positivist science, they have sought to understand why we perceive our environments as we do. But their methods and aims have not been exclusively positivistic (Gold 1980) because they have been dealing essentially with irrational, emotional and non-predictable man. Thus, as Jeans (1974) sees it, behavioural geography posits that the 'real' environment is seen 'through a cultural filter made up of attitudes, limits set by observation techniques and past experience' (see Figure 1). To understand decision making we need to study that filter, which is part of the means whereby human consciousness structures reality. But at the same time, in this view there *is* a 'reality' which is separated from consciousness. The two are not entirely a function of, or fused with, one another as would be the case from a phenomenological viewpoint or from that of organic systems advocates like Capra, who base their ideas on a fusion of mysticism with the most recent science.

To conclude, the determinist/free-will philosophical spectrum is really fundamental in understanding the roots of environmentalism for three reasons.

First of all, when we think about the man-nature relationsip, we all express ideas (whether we realise it or not) which relate to this spectrum. These ideas say to what extent we regard ourselves as part of nature or separate from it, and to what extent we think that we can dominate and exploit it. Secondly our philosophy expresses how much we think that we are at the mercy of natural or other 'external' forces, and therefore it governs to what extent we think we can work for a different (more ecologically and socially harmonious) future via social change. Third, it sheds much light on the question of whether the messages we get about nature come from an 'objective' source. The weight of opinion, from the free will end of the spectrum,is that Cartesian objectivity is impossible.

5.3 SCIENCE AND IDEOLOGY IN THE PAST 200 YEARS

We now examine the idea of objectivity in a slightly different sense, that is in the sense of 'bias'. Even assuming that there *was* an external nature of 'hard' measurable 'facts', we would have to ask whether scientists were likely to investigate and represent those facts in a way which did not relate to their ideological preconceptions and/or the preconceptions of groups from which scientists draw their ideal and material sustenance.

Modern science historians and sociologists increasingly tell us that frequently scientists are not likely to be neutral in this way and that in a sense (perhaps unconsciously) conclusions are *first* drawn about the subjects and outcomes of scientific inquiry and *then* science is used to provide data, theories and 'laws' which support and give legitimacy to those conclusions. Marxist commentators might go further and tell us that such pre-conclusions tend to serve the interests of specific groups in society. These groups - rather than the scientist alone - begin the process of 'biassing' science by having a large say in *what problems* are considered to be important and worth investigating in the first place. They continue by influencing *what data* out of a multitude are significant. Science is therefore used to support the ideologies of particular groups; thus it is not 'objective' - it sustains vested interests. Furthermore, if, as a result of these ideologies, the interests of some other groups suffer, the sufferers will be more inclined to bear their burdens uncomplainingly if it can be shown that they arise from the operation of 'scientific laws'. For scientific laws are inevitable and unavoidable - they are universal generalisations which are 'above politics' and political action. Like science itself, scientific laws are seen as 'value free', so they cannot be changed just because some people think that they breed undesirable results. Social injustice can thus become more acceptable, bearable and legitimate if it can be shown to result from laws of nature - if it is 'natural'. Cotgrove sums up this modern view of science as ideology:

It is not, of course, claimed that science is merely ideology, but only that ideology is an important element. There has been a growing realisation of the basically subjective and ideological aspect of science which stands in sharp contrast to the simplistic positivistic pretensions that there are plain facts waiting to be objectively observed and reduced to law-like statements. The claim that science is exclusively an objective enquiry and in this sense, divorced from politics, is therefore becoming much more difficult to sustain. Not only, it is argued, is science used to justify particular models of political and economic behaviour (scientism) but the subjectivity in science includes models and assumptions about nature which resonate with the political perspective of scientists. Thus, as Reich points out, appeals to nature and human nature are used to shift the problem from the social to the biological sphere, where nothing can be done about it, and are in this sense essentially conservative. Such charges are usually demonstrated in the field of biology, where there is a long tradition (Adam Smith, Malthus, Spencer) of using science as a rationalisation and justification for particular economic and political models - notably the inevitability of inequality, hierarchy and competition. (Cotgrove 1975 pp71-2).

We see that there are several aspects of Cotgrove's argument which might be said to apply to the role of science and scientists in environmental issues. These aspects are (a) that science is not divorced from 'politics' (b) that it is essentially conservative (c) that it is used to justify particular economic and political models of society. We shall go on to examine how they could apply to modern environmental issues, but first, in keeping with the approach of this book we shall take an historical perspective and see these ideological functions of science and scientists in relation to some examples of 18th- and 19th-century environmental questions. One is concerned with a specific case of pollution which illustrates that science was not divorced from politics 200 years ago; the second illustrates, through Malthus, how science was used openly in support of a conservative ideology; the third discusses social Darwinism as an example of the use of science to justify particular economic and political social models.

5.3.1 A Case of Pollution in Rouen
The theme of science and scientists in support of vested economic interests, and therefore not divorced from politics, is illustrated in John Perkins' study of one of the earliest recorded examples of a modern pollution case. It took place in Rouen between 1785 and 1815 and its history helps to show how the characteristics of 'classical' science developed in industrialising Europe and how that science rose to its present pre-eminent position in our 'cultural filter' (Perkins 1984). During this period, science became specialised and professionalised and incorporated into the educational system. And, most important here, the *scientific expert* emerged, with his access to difficult knowledge and to the political decision makers in the state bureaucracy. He became a respected arbiter in cases where the interests of man and nature clashed.

In this particular case he was called in to pronounce on the effects of the noxious gases of the oxides of nitrogen and hydrochloric acid which were emitted respectively in sulphuric acid and soda manufacturing in the southern suburbs of the city of Rouen. This industry boomed in the first decade of the 19th century, and was transformed from an artisan base to a large-scale capital-intensive activity through the efforts of new industrial entrepreneurs, who came in to take advantage of the huge profit opportunities. The subsequent collapse of demand around 1810 was a measure of the financial precariousness of the industry, of which the manufacturers were always sensible. They were therefore concerned to maximise profit while it was available, and they opposed any regulations and restrictions on production. But this *economic* interest ran counter to the *economic* interest of the opponents of the industry. These were principally from among the older established commercial bourgeoisie, who lived within the old city and made money from renting property - either for housing or for horticulture - in the suburbs where the new industry came to be located. That area was particularly prone to temperature inversions, and these people were frequently subjected to acid fogs.

By petitioning and other means the opponents claimed that such pollution was damaging their property and reducing its value. In support, they pointed to visible evidence of damage to plants and animals, and to human health. The manufacturers, however, claimed that there *was no* damage, or, if there were, that it was not related to their factories. The chief administrator called in experts for advice. He turned to a new type of figure who was beginning to appear in many European countries at this time, the professional scientist, who earned his living by practising science, teaching and popularising it. Those he called on were chemists - local teachers of the science - whose intellectual affiliations were with new national developments in scientific knowledge - above all with the revolution in chemical theory associated with Antoine Lavoisier (1743-1794). The administrator ignored those who had been the established experts before the French Revolution, the apothecaries and physicians. As members of guilds their social and cultural affiliations had been with the city's commercial bourgeoisie. Unlike the new professional chemists they were not closely identified with the new developments in chemical theory.

When they came to investigate the charges of pollution the new professional scientists did not, however, use the supposed scientific method. They did not try to observe the claimed damage, to substantiate whether or not it had taken place, or whether it had resulted from pollution. Instead, they argued from the position of their privileged access to what was accepted as chemical knowledge - from their position as representatives of that knowledge - in order to support the manufacturers and to repudiate their opponents. They said that their theoretical understanding of the manufacturing processes showed that acid gases *could not* have been produced. And when they had to agree that there was some damage, for example after being shown dead vegetation, they again argued from theory, to assert that the cause could not have been acid gases because even if they were present they were known to be harmless, indeed, even beneficial. They attributed the effects to speculative causes, such as soil exhaustion, adverse weather, insect damage or bad husbandry. The opponents' reaction was to reject the findings of these 'experts' and call for the use of the pre-Revolutionary locally-based traditional experts, and to petition that the authorities should come and see for themselves. But their demand for an empirical, observational, 'common-sense' approach was rejected in favour of the new scientists' 'rational' deductive and theoretical approach. The opponents then shifted ground and accused the experts of not being objective, but of displaying a bias which stemmed from their vested interests. And indeed it transpired that some of the chemists called in by the Prefect were, themselves, actually involved in financing or carrying out the new manufactures - they had industrial interests which identified them economically with one side. However, the administration refused to concede that their experts could be thus tainted, and by doing so - by asserting that any vested interest was irrelevant - they implied that the new scientists were agents for impersonal, value-free knowledge. When the opponents in response claimed that *theory*, as well as observation, supported their case, the scientific experts did not attempt to refute their theoretical arguments; instead they ridiculed them as out-moded, fanciful and irrational. Only the professional, they implied, has access to correct knowledge.

Modern parallels will strike the observer of environmental issues in the 1970s. Discrediting the opposition has been a feature of nuclear controversies (Del Sesto 1980, Nelkin 1975). But the vested interests of nuclear experts have been recognised, as witness the identifications made in a Granada TV programme in 1983, of links between the Central Electricity Generating Board chief Walter Marshall and the Westinghouse Corporation, manufacturer of the pressurised water nuclear reactors - for which he was campaigning as a major means by which the CEGB should produce its future supply (see section 6.5).

From the 19th-century example, Perkins shows us how the germs of ecocentric and technocentric attitudes emerged. The opponents in the case emphasised empirical knowledge and 'common-sense' attitudes, using language which was dubbed as emotional and exaggerated. The manufacturers maintained that *they* were, by contrast, reflective, rational and objective - they appropriated for themselves what we have come to regard as legitimate scientific knowledge and its language and style.

5.3.2 Malthus and Ideology

In the above example science and scientists were used to advance the interests of a new group in society - a group of industrial capitalist entrepreneurs - at the expense of the more established landed classes. That this was not uncommon by this time has also been demonstrated by Thackray (1974) who showed how the emerging industrialists of Manchester at the turn of the 18th and 19th centuries participated in science and scientific institutions to advance their claims for social recognition. But, as Thackray also showed, it was not long before science became dissociated from new radical groups and ideas, and instead became enlisted in the interests of preserving the social, political and economic *status quo*, thus demonstrating clearly that it is not neutral within the context of society. Such conservatism, as Cotgrove says, is what commentators who see it as essentially a social activity now more usually associated with science. And one of the most influential initiators of this tradition must have been Thomas Malthus.

The charge against Malthus is that he provided, in the guise of an apparently objective formulation of scientific laws ('principles' of population - see 4.1.2 and 4.1.3) a rationale for not supporting the poor through state intervention in the form of the Poor Laws. He justified the continuation of gross inequalities of wealth distribution, and of the political and economic *status quo*, including the ownership of land predominantly by the aristocracy. He put the blame for the plight of the poor *on* the poor and exonerated the ruling classes from responsibility. And in so doing he essentially preached against social reform and revolution. Perelman (1979) sums up the charge:

> The concept of scarcity as it appeared in the ideological struggle about the poor laws was very crude, so Malthus' simplistic formulation served admirably as a political weapon. Malthus proved to the satisfaction of the ruling classes that they had no responsibility for the existing state of affairs. They were not about to raise questions about subjects such as the effect of private property on the availability of resources: it was enough for them that Malthus showed that '...the real cause of the continued depression and poverty of the lower classes of society was the growth of population'. Who among the ruling classes would question the doctrine that the road to salvation lay not in furthering the struggle between classes but in eliminating the lust between sexes? The Reverend Malthus conveniently absolved all members of the parish of capital. The poor are, Malthus told his contemporaries, 'the arbiters of their own destiny; and what others can do for them is like dust in the balance compared to what they can do for themselves' (Perelman 1979 p81).

There are those who argue that Malthus has been misjudged, and that 'his political activism was devoted to increasing the educational and social welfare provisions for the poor and working classes of England' (Schnaiberg 1980). It is not our purpose here to debate which side is 'correct'; for the present we need to examine how it is that one side, at least, *can* argue for his theory being essentially ideological.

To appreciate their argument, one must be aware of some facts about the social climate within which Malthus wrote, at the beginning of the 19th century and later. First, it appears to have been a period of population increase in Europe, though many, such as Cobbett (1830 'from Kensington to Uphusband') strenuously denied this. Secondly the growth of the factory system was leading to the population drift from country to town, with a concomitant phenomenal rise in the urban population and urban squalor and deprivation. Thirdly, there was the rise of a new class of industrial entrepreneurs who were challenging the established order - the aristocracy, which class Malthus is usually held to serve - and its ownership of the means of production, especially land. Fourth, there was a revolutionary climate in Europe generally - as Perelman puts it 'Malthus was writing when almost all of well-to-do England lived in constant horror of the egalitarian principles of the

French Revolution'.

The question we have to consider is, were these changes leading to 'overpopulation', in the sense that people were breeding at a rate which outstripped the growth in food supplies? Perelman's view is that they were not, and that 'the revolutionary changes in English agriculture and industry were eliminating traditional forms of employment faster than new industries could create alternative employment, producing an apparent population surplus and attendant poverty'. This created a burden of poor relief which 'seemed an unnecessary expense to the bourgeoisie and also appeared to reduce the necessity [for the poor] to accept employment in the "dark satanic mills" of the industrial revolution'. Malthus' argument was that to give institutional poor relief (large-scale state relief as opposed to selective private charity to the 'deserving') would tamper with the *natural law*. It would hinder the operation of positive checks (famine and disease) to population growth, and, unfortunately, one could not anticipate that in their absence the preventive checks of moral restraint etc. would be brought into play to limit population. Here, Malthus's class bias emerged. He was up against the paradox that the relationship between financial ability to rear children and the number of children actually born was an *inverse* one, for, as he saw, the aristocracy did not have as many children as their means would theoretically have supported. Yet at the same time he was postulating that increased affluence led to *increased* population numbers. To substantiate this paradox he appears to have argued literally that there was one law for the rich and one for the poor. Petersen (1979) suggests that in this he was much influenced by Adam Smith, who wrote in *The Wealth of Nations* 'Every species of animals naturally multiplies in proportion to the means of their subsistence, and no species can ever multiply beyond it. But in a civilised society it is only among the inferior ranks of people that scantness of subsistence can set limits to the further multiplication of the human species - and it can do it in no other way than by destroying a great part of the children...'

From this, Malthus seemed to advocate that while the aristocracy were responsible enough to practise moral restraint - not transforming increased wealth into more children - the working class were not. This was because they lacked sufficient education and enlightenment (e.g. about the 'principle of population'). He was pessimistic that they ever would acquire it - perhaps because he saw that there were political forces opposed to the education of the poor - and so came inexorably to the conclusion that as preventive checks would not operate with the poor, positive checks would have to be given free rein. Otherwise the population would grow, increasing poverty and misery. So however well-intentioned the rich might be, they should not offer widespread poor relief, which only exacerbated the poor's plight. The solution to the latter's problem lay in their own hands, not in general social reform. It lay in moral restraint, such as that already practised by the better-educated and more affluent.

Thus, because of the inexorable operation of scientific principles, the plight of the poor could not be alleviated by institutional means like poor laws. According to Glacken (1967) Malthus wrote to Godwin that although 'Human institutions appear to be...the obvious and obtrusive causes of much mischief to society, they are, in reality, light and superficial in comparison with those deeper seated causes of evil which result from the laws of nature and the passions of mankind'. Chase (1980) regards this as sheer 'scientism', i.e. 'Using the language, symbols, findings and other attributes of the legitimate sciences to advance unproven preconceptions and dogmas'. Chase concludes that Malthus was thus the forerunner of 'scientific racism', in that he sought to provide scientific excuses for not undertaking any measures to protect populations from preventable diseases and disorders (and, we must include, poverty). Chase regards Malthus' presentation of his natural law of population in the form of a 'mathematical statement of God's will' as the first major theory of the human inferiority of the 'lower and middling classes' to be presented in the language of science. The function of scientific racism is to preserve poverty, because it is the source of cheap labour (it is, as Malthus put it, 'The necessary stimulus to industry'), and to prevent the passage of legislation to improve people's lot.

For this reason, and because philosophically it denied man's ability to take charge of his own destiny and improve society through reform (it was determinist), Marx objected to Malthus's thesis as a 'libel on the human race'. Not only did Malthus put 'poverty's base in nature' rather than in society, he also advocated that his Principle should be 'generally circulated' as a means of defusing revolutionary fervour. The reasoning was that if you know that your poor position results from the operation of universal laws which no amount of social reform can resist, then you will be more inclined to accept your position. The position is *legitimated*:

> there is one right which man has generally been thought to possess, which I am confident he neither does nor can possess - a right to subsistence when his labour will not fairly purchase it. Our laws say that he has this right, and bind the society to furnish employment and food to those who cannot get them in the regular market, but in so doing they attempt to reverse the laws of nature...if the great truths on these subjects were more generally circulated, and the lower classes of people could be convinced that by the laws of nature, independently of any particular institutions, except the great one of property, which is absolutely necessary in order to attain any considerable produce, no person has any claim of *right* on society for subsistence if his labour will not purchase it, *the greatest part of the mischievous declamation on the unjust institutions of society would fall powerless* to the ground.

Scientific truth, then, is in this case counter revolutionary, and

> If these causes were properly explained to them [the poor]... discontent and irritation among the lower classes of people would show themselves much less frequently than at present... The efforts of turbulent and discontented men in the middle classes of society might safely be disregarded if the poor were so far enlightened... (Malthus 1872 p191, Book4, Chapter 6, emphases, except the first, added)

This naked appeal to vested interests, and the use of science to justify it, is confirmed in the concluding chapter, 14, of Book 4 of the seventh edition of the *Essay*. The chapter is on 'Our Rational Expectations Respecting the Future Improvement of Society'. Malthus makes it clear that his work should not be interpreted as an argument for social reform:

> It is less the object of the present work to propose new plans of improving society than to inculcate the necessity of resting contented with that mode of improvement which has already in part been acted upon as dictated by the course of nature... the limited good which it is sometimes in our power to effect is often lost by attempting too much,...

Indeed, the burdens placed on the aristocracy should be lessened:

> The gradual abolition of the poor-laws has already often been proposed in consequence of the practical evils which have been found to flow from them, and the danger of their becoming a weight absolutely intolerable on the landed property of the kingdom.

The rich should not waste their efforts in trying to do the impossible (large scale poor relief) and should concentrate on encouraging the poor to practise restraint. 'This knowledge, by tending to prevent the rich from destroying the good effects of their own exertions, and wasting their efforts in a direction where success is unattainable, would confine their attention to the proper objects, and thus enable them to do more good.' Scientific laws mean that social reform is bound to be unable to eliminate poverty, and if only the poor knew of such truths, then they would accept their political fate, be grateful for what they were given and be disinclined to start a revolution:

> ...that the principal and most permanent cause of poverty has little or no *direct* relation to forms of government, or the unequal division of property; and that, as the rich do not in reality possess the *power* of finding employment and maintenance for the poor the poor cannot, in the nature of things, possess the *right* to demand them; are important truths flowing from the principle of population, which, when properly explained, would by no means be above the most ordinary comprehensions. And it is evident that every man in the lower classes of society who became acquainted with these truths, would be disposed to bear the distresses in which he might be involved with more patience; would feel less discontent and irritation at the government and the higher classes of society, on account of his poverty; would be on all occasions less disposed to insubordination and turbulence; and if he received assistance, either from any public institution or the hand of private charity, he would receive it with more thankfulness, and more justly appreciate its value.
> If these truths were by degrees more generally known... the lower classes of people, as a body, would become more peaceable and orderly, would be less inclined to tumultuous proceedings in seasons of scarcity, and would at all times be less influenced by inflammatory and seditious publications, from knowing how little the price of labour and the means of supporting a family depend upon a revolution. The mere knowledge of these truths... would still have a most beneficial effect on their conduct in a political light; and undoubtedly, one of the most valuable of these effects would be the power

that would result to the higher and middle classes of society, of gradually improving their governments, without the apprehension of those revolutionary excesses, the fear of which, at present, threatens to deprive Europe even of that degree of liberty which she had before experienced to be practicable,... (Malthus 1872 pp260-261).

5.3.3 Social Darwinism and the Justification of Politico-Economic Models of Society

Darwin saw his thesis about the struggle for existence which followed from the plenitude of nature as 'the doctrine of Malthus applied in manifold force to the whole animal and vegetable kingdom' (Darwin 1885 p59), except that in these kingdoms the exercise of preventive checks was impossible. In this sense his picture was, therefore, an even more deterministic one. If one remembers this, it is easy to see that the principles of Darwinism can create bitter hostility from some quarters when they are applied to the *behaviour of man*. Malthus' determinism - his 'libel on the human race' - becomes even more of a libel after it has been intensified through being applied to the vegetable and animal kingdoms, and then taken and applied *back* to human society, which is what happens in Social Darwinism. In this doctrine, human behaviour at large - not just in respect of reproduction and population size - is compared to models of animal behaviour, and the 'essentially competitive nature' of man is held to derive directly from Darwin's model of populations growing to exceed available resources, which leads to a struggle between groups and individuals in which the fittest survive, ensuring the sophistication of the species. However, Social Darwinism was not a doctrine which Darwin himself proposed. It was others, such as Herbert Spencer, who took Darwin's ideas and applied them to society.

There can be little doubt that, despite the influences of philosophies of free will, and of egalitarian movements for social reform in the West, the ideas of Social Darwinism continue to be profoundly influential today. Schumacher (1973 p71) holds them to be part of the 'hidden curriculum' in education, in which we learn without realising it to regard man as being inherently - that is *by nature* - competitive. And we accept that the fittest should survive the struggle while the weakest should not, in an inexorable and desirable evolutionary process wherein some groups and individuals dominate and other are dominated. In Social Darwinism, then, the analogy is made between what apparently happens inevitably in 'nature' - as defined by the *scientific principles* elucidated by Darwin - and what happens in human society. When this happens it becomes easy to accept animal-like behaviour from men, and to see it as inevitable and as justifiable as any other phenomenon which is described and understood via the medium of scientific laws. And when we examine some of the laws of animal behaviour which Darwin described, we can see just what it is which is being justified through the application of Darwin's science. We can see, that is, what Albury and Schwartz (1982) call the ideology behind scientific theory, and how with Darwin's science the 'locus of inequality' between men moved from 'God's will' to natural/biological explanations:

...we should remember how essential it is in a flock of white sheep to destroy every lamb with the faintest trace of black (Darwin 1885 p77)

...to my imagination it is far more satisfactory to look at such instincts as the young cuckoo ejecting its foster brothers - ants making slaves... not as specially endowed or created instincts but as small consequences of one general law, leading to the advancement of all organic beings, namely, multiply, vary, let the strongest live and the weakest die. (*idem* p219)

Natural selections in each well-stocked country, must act chiefly through the competition of the inhabitants one with another...Hence the inhabitants of one country, generally the smaller one, will often yield, as we see they do yield, to the inhabitants of another and generally larger country. For in the larger country there will have existed more individuals, and more diversified forms, and the competition will have been severer and thus the standard of perfection will have been rendered higher. (*idem* p184)

As natural selection acts by competition, it adapts the inhabitants of each country only in relation to the degree of perfection of their associates; so that we need feel no surprise at the inhabitants of any one country, although on the ordinary view supposed to have been specially created and adapted for that country, being beaten and supplanted by the naturalised productions from another land (*idem* p424)

...The mental quality of our domestic animals vary, and...the variations are inherited... (*idem* p218)

In these quotations we can see that in the natural world the weakest can and should 'go to the wall', slavery is justified, as is territorial aggression, and that not only are physical characteristics inherited, so are mental characteristics. Clearly such a perspective applied to the *human* world can be used to justify ideological positions involving territorial aggrandisement, the *lebensraum* thesis, the analogy between states and organisms in which the former display organic lusts, the idea that some groups are genetically superior to others (including intellectual superiority), and slavery. If one tries to argue that man is different, and that therefore Darwin's science does not justify 'natural' behaviour patterns in society, Darwin's science can be invoked again. For it heavily implies that man is a part of nature, from which he developed. He is part of the 'web of life', or, as ecocentrics now say, a component of natural ecosystems. If he is part of nature then it is not easy to see how he can be freed from such natural law as that of the survival of the fittest.

One of the most familiar spheres in which Social Darwinism is manifest is that of economics, where the whole paraphernalia of competition, struggle, dog-eat-dog is to be discovered in the world of 'free enterprise' capitalism. The behaviour of the business and industrial world is often morally repugnant to those who pause to stop and reflect on it (we are familiar with such examples as the arms trade, or the drive to sell dried milk or strong cigarettes to the Third World). But it usually can be 'understood' i.e. excused and legitimised, if it is seen as the result of the operation of scientific 'economic' laws, analogous with those natural laws discovered by Darwin. Albury and Schwartz (1982) discuss the current popularity of ethology - the study of animal behaviour - and they suggest that this study shows that animals behave in the same was as do humans under capitalism - they are aggressive, possessive and competitive. Researchers conclude that if animals do it such behaviour is *universal*, and if it is universal then is is excusable and acceptable (cf. the popularity of Desmond Morris' *Naked Ape* and of the kind of view which accepts that 'Recent ethological studies on birds, apes, and other animals have unmistakably demonstrated how much there is in common between the behaviour of young animals and the human infant' - Nicholson 1970).

The critics of Social Darwinism often stress this link between the 'mode of production' i.e. capitalism, and the predominant ideas in society (i.e. Darwinistic ideas). They advocate that such a link is not merely fortuitous, but that it arises because Darwin's science is inherently ideological - being a transference from society to nature and back again of capitalist ideology - after which process the ideology becomes respectable:

> The whole Darwinian theory of the struggle for life is simply the transference from society to organic nature...of the bourgeois economic theory of competition. Once this feat has been accomplished... it is very easy to transfer these theories back again from the natural world to the history of society, and altogether too naive to maintain that thereby these assertions have been proved as eternal natural laws of society (Engels 1963).

> It is noteworthy that Darwin rediscovers in animals and plants his own English society with its social division of labour, competition, the opening up of fresh markets, invention and the Malthusian struggle for existence. This is the *bellum omnium contra omnes* of Hobbes... (Letter from Marx to Engels dated 18 June 1862.)

In this argument it is not suggested that scientists deliberately and premeditatedly obscure and distort what they see for ideological ends. Rather, as Albury and Schwartz point out in respect of the sociobiologist E.O. Wilson, the scientist reports what he sees in animal behaviour (including aggression, genocide, male dominance, militarism, spite, territoriality). But *what* he sees - in the sense of deciding what is 'significant' - is strongly influenced by his 'perception framework': that is, his whole consciousness as determined by his educational and cultural experiences, or his cultural filter. It follows that if this filter is dominated by a particular ideology then what is perceived through it as significant will in effect be tailored to that ideology. If we understand this line of criticism of Social Darwinism, then we are some way towards understanding the Marxist dictum that it is not the consciousness (i.e. cultural filter) of men which determines their being (i.e. their economic and material existence) but their social being which determines consciousness.

5.4 SCIENCE AND TECHNOLOGY AND THEIR 20th CENTURY IDEOLOGICAL CONTEXT

We have explored the assertion that science is a social activity, and that science and scientists are essentially influenced in what they do by their social context, and our examples so far have been drawn from the 18th and 19th centuries. We can now carry the argument forward in time, and also see how some writers have developed various aspects of it. These aspects are: (a) that 20th-century science, in the form of technical expertise and the appeal to universal laws, is used more strongly than ever in the legitimation process; (b) that the scientific research which is or is not done - and the technological developments which stem from it - is essentially *selected* according to whether or not it supports the ideologies and purposes of particular groups; (c) that these practices can be seen as tending to increase the amount of social and economic control exerted by select groups over the lives of ordinary citizens.

(a) Legitimation

Nelkin (1975) has examined the role of technical expertise in influencing decisions on two environmental issues - the proposed siting of a nuclear power station by Cayuga Lake, New York State, and a proposed new runway for Logan International Airport in Boston, Mass. She quotes Ezrahi (1971) who says 'The capacity of science to authorise and certify facts and pictures of reality is a potent source of political influence'. Such influence, she maintains, was sought by developers and by their opponents in her case studies by using 'scientific experts'. On the developers' side they would try to show that the proposed schemes were urgent for economic (and if possible environmental) reasons. They would produce complicated technical reports, requiring considerable expertise to understand, which would play down adverse environmental effects, and support the notion that the proposed scheme was the only viable one from the economic and technical point of view. Evidence from this source would aim to restrict discussion to questions of *how* to achieve aims like meeting an assumed increase in demand for electricity or for air travel. It would not encourage questioning of whether these aims *should* be achieved. On the other hand, the 'experts' on the other side would attempt to challenge the assumptions, both that there would be increased demand, and that it should be met at the expense of some environmental impact. Nelkin points out that the available data were used by each side to make different predictions about the impact of the proposed development. In this sense the data - the 'facts' - were irrelevant; 'as for the details of the technical dispute, they had little bearing on the case'. Nelkin's conclusions reinforce the unimportance of 'objective facts', for she believes that in such cases the way clients use their experts essentially 'embodies their subjective construction of reality', and that the extent to which expertise is accepted depends less on its validity and the competence of the expert than on the extent to which it 'reinforces existing positions'. So the main point in clients using scientific experts is to achieve a political influence which derives from the widespread perception by the public and media that experts are rational, objective and making technical and non-political points.

That is, *political influence* is sought by attempting to convey the impression that the case which is being made is a *non-political one*.

This 'non-political' theme may be examined further. What it means here is that for a development to be regarded as 'non-political' it must be seen to be in the interests of *society as a whole* rather than of any narrow section of society. To appeal to the universal good, or to the 'national need' tends to be regarded popularly as 'non-political' and therefore more worthy of support than a 'narrow' political end. Whether we agree with this popular view or not, we can recognise that it is widely held, and is perhaps based on the rather low level of political awareness of many people and on the widespread mistrust in which the professional politician is generally held. So in order to gain wide acceptance - that is, in order to be legitimate in the eyes of the majority - one must *universalise*, i.e. appear to appeal to a broad rather than to a narrow interest, like Bacon's man of science (see section 2.4).

Del Sesto (1980) has illustrated this in his discussion of the ideologies of each side in the nuclear power debate (see Figure 3). In Congressional hearings on The Status of Nuclear Reactor Safety held in 1973-4, the pro-nuclear lobby played down its own potential benefit if a pro-nuclear policy were to be accepted by Congress. Rather it emphasised the 'convincing central message that nuclear energy was good for society' and that it ...'ultimately could benefit human welfare by improving the standard of living and increase the rate of economic growth' (for all). These themes, together with the achievement of energy self sufficiency for the US in the face of the (external) Arab Oil threat, are said to have 'continually surfaced in the hearings'. The General Manager of the Boiling Water Projects Division of General Electric, a leading manufacturer of nuclear plants, emphasised that to support his interests was equivalent to supporting general social progress - 'if we are to maintain our present standard of living, as well as make it available to all segments of our society', he said, then more electricity would be needed (Del Sesto 1980 p44). This was predictably followed by references to 'urgency' and 'rapidly escalating costs of fossil fuels' in which 'pro-nuclear witnesses assumed that increasing the supply of electricity and improving the standard of living were causally related', and 'pro-nuclear technology was dominated by [this] faith in science and technology'. As Nelkin also did in her case studies, Del Sesto referred to the way that pro-nuclear witnesses saw the use of technical expertise as a way to interpet and resolve issues - 80% of their testimony was 'couched in terms of technical facts-and-figures' as compared to 17% for the anti-nuclear groups. Although both sides in the nuclear controversy do attempt to legitimate their arguments via this kind of scientific evidence, the anti-nuclear-power lobby tends to resort to this tactic less - partly for financial reasons (the public inquiry into the proposed pressurised water reactor at Sizewell in Suffolk illustrated in 1983 that the financial resources of opposing groups such as Friends of the Earth were far inferior to those of government-backed proposing groups such as the CEGB, which spent £10m on assembling its evidence) and partly because of its mixed scientific and non-scientific philosophical and ideological roots.

The process of legitimation by appeal to apparently universal scientific principles is, it will be clear from the above, extended to the economic sphere. Economic scientism is a familiar theme to observers of the proponents of 'monetarism'. In this ideology, where the operation of the market place is deemed sacrosanct in determining public policy, particular economic perspectives are elevated to the status of scientific 'laws' comparable with those of physics, as Prime Minister Margaret Thatcher demonstrated: 'There is this [incorrect] belief that the laws of economic gravity can somehow be suspended in our favour' (Conservative Party Conference, 1981). The benefits of doing this if you are a monetarist are the same as those which Malthus perceived in formulating his principles of population. When laws are influencing your life they tend to be incontrovertible, and life is *determined for you* in this sense. You *cannot* pay yourself more than you earn, you *cannot* increase public spending without economic growth, you *cannot* be competitive with 'over-manning', and you *cannot* survive without competing, in just the same way as an apple *cannot* fall upwards when it is detached from its tree. And furthermore, it is not just *you* who must obey the law, it is everyone who must obey the universal truth - in space and time. No amount of attempts at socialism or other social reform can apparently overcome the power of the universally iron laws of economics, so such attempts had better not be made lest they prove 'counter productive', or 'unhelpful', or - worst of all - 'irresponsible'.

In just the same way, you cannot deny 'progress'. And if progress is equated with high technological development, then you cannot deny that. Those who do deny it are somehow negative and nihilist, and equatable with those well-known machine breakers, the Luddites, who are written off as an expression of working class immaturity, in which the working class critics of technology are stigmatised as advocating a return to an 'aristocratic golden past' (Albury and Schwartz 1982). Thus, if you oppose a microprocessor and other electronic advances which replace jobs then you are being Luddite, and opposing the scientifically inevitable, the technologically inexorable. This is a powerful message, and IBM has not been slow to broadcast it (see Figure 7). Its success seems to be borne out by the air of resignation which all sides of industry - including the unions (see Webster and Robins 1982) - exhibit in the face of de-manning and rising unemployment abetted by technological innovation. (A Marplan poll in November 1982 reported that 57% of all voters thought that no party could significantly reduce unemployment levels - the matter had thus become 'non-political'. And the processes of legitimation and universalisation had been successful enough to persuade 4 in 10 that the monetarist Government was not to blame for high unemployment, which would be with us for many years to come; *Guardian* 23 November 1982). Indeed, resignation is quickly extended to a form of social Darwinistic determinism, as the following quote, resonant with biological analogy, shows: 'Britain needs five million unemployed...in order to promote a healthy economy, writes editor Arthur Seldon in the *Journal of the Institute of Economic Affairs* ... Unemployment at the level

You can't stop time by smashing clocks.

Between the years 1811 and 1816, a band of textile workers had just the answer to the threat of technology.

They literally threw spanners into the works.

And smashed up the new machinery which they blamed for their unemployment and distress.

If this attitude had prevailed, weaving would still be a cottage industry.

Ploughs would never have exceeded 4 horse-power.

The steam engine would have lost out to the cart driver.

And Britain would never have become the economic power it was in the nineteenth century.

Yet the action of the Luddites carries a very instructive lesson: it's not progress itself which is the threat, but the way we adapt to it.

For without technology, a nation's progress would undoubtedly falter.

Machines bring down the cost of production.

Which in turn either creates greater profit for reinvestment, or holds down the costs of the product, so providing greater purchasing power for the pound.

The result is greater wealth – the ideal climate for increased employment.

And machines that relieve man of the tasks that limit his personal fulfilment.

Smashing the clocks might destroy the mechanism of progress.

But it will never delay tomorrow.

IBM

IBM UNITED KINGDOM LIMITED P.O. BOX 41 NORTH HARBOUR PORTSMOUTH PO6 3AU

FIGURE 7   **Technology Equals Progress: Newspaper Advertisement 1979 (By Permission of IBM United Kingdom)**

proposed is "inevitable, necessary and desirable... Progress must create unemployment. We shall have to see much more unemployment before the British economy is freed of its deadwood to grow fresh shoots"' (Quoted in the magazine *Public Service*, September 1982).

(b) Sustaining Vested Interests
Albury and Schwartz go further in discussing the ideology of science and technology. They support Luddism, seeing it not as a regressive force, but as a means of attaining workers' control over science and technology. Once this control is attained, one cannot assume that it will be over the *same* scientific and technological developments as those which are now in vogue. For to say (as some on the left do) that there is nothing inherently ideological in the technology itself, and that under common ownership any technology (such as nuclear power) can be benign, is too simple a position for Albury and Schwartz. They maintain that the myth that the products of research and development are basically positive, standing above social conflict and social divisions and acting for humanity as a whole, originated in a specific historical period - the 19th century - and was identified with the highest development of a specific economic system, that of industrial capitalism. They believe that this identification still holds today, and that the *very problems which have been set for scientists and technologists to solve have been essentially the problems of the owners and controllers of industry* and its allied concerns, rather than the problems of society as a whole. Thus, what is in the 'national interest' is decided for the many by the few, and in favour of the ideology of a particular class. In this process, the Government's role is as a major funder of scientific research in defence of the capital owning classes.

Nelkin (1975) makes much the same point: 'the viability of bureaucracies depends on the control and monopoly of knowledge in a specific area', and she follows it up with the observation that scientists today are traditionally involved in Government entrepreneurial areas (e.g. space research) or policy areas (defence). They have minimal participation in policy areas with wealth redistributive implications - that is to say, they support, wittingly or unwittingly, the aims and goals of a specific group, to the detriment of all of society. (Though, of course, in the way of legitimation, representatives of that group, such as the Confederation of British Industry will argue that what is good for *it* is good for all.) It follows from all this that if control over scientific and technological research and development changes, then the type of products which it provides will also change, to those which advance the interests of the new controlling group (e.g. if they were ecocentrics, to job creation rather than destruction, to 'soft' energy, to 'organic' farming etc.). The present control tends to lead to overproduction of some products (like arms) existing alongside unmet social needs, because the former meet profit-creating criteria while the latter (education, social services) often do not. And, as Robins and Webster (1983) point out, new technologies like 'information' technology, have 'everything to do with the continuation of the old order'.

Albury and Schwartz trace the development of several different areas of science and technology, in order to demonstrate how such advances were serving the particular needs and ideologies of capitalists. Thus, capitalism developed by expanding its markets. Such expansion came about by several means, including the territorial extension of trading. When this happened, two things followed. First, fast means of communication had to be developed between the industrial heartland and its outposts: thus a large telecommunications industry (represented by companies like Siemens AG) was founded and encouraged by the interests of capital. Second, these outposts had to be defended against those who would limit their right to 'free' trade with the heartland, so a 'defence' industry was fostered. Now, of course, the two industries serve each other as well as the interests of other industrial corporations.

Another way of expanding markets lies in product innovation, and it can be clearly seen that the microelectronics industry has served this particular need. There has in recent years been a plethora of luxury consumer goods incorporating the microprocessor, a development of the telecommunications industry. To accumulate capital, the capitalist can also try to replace his workforce by machines whenever possible. Machines may or may not be cheaper than people to 'buy', but their running costs tend to be lower. Thus machines do not cause industrial disruption and productivity loss in demanding better pay and conditions. They do not oppose their own redundancy. And they also mean that the remaining jobs done by people often become de-skilled, and un- or semi-skilled labour is cheaper and less burdensome to the employer than skilled labour. Here again, research into robots, computers and word processors has been prioritised by industry and government - 1982 was designated in Britain 'Information Technology Year' and special grants were available to universities and polytechnics for research into microprocessor applications (but not for research into the social consequences of such applications).

One powerful private funding body is the Rockefeller Institute, and Albury and Schwartz document the involvement of this organisation in funding developments which would ultimately help a 'Green Revolution' in Third World countries. In particular, a conscious decision was taken in 1932 to channel funds into biological research. This was to lead to the accelerated development of plant genetics, and to wheat and rice plants of uniform height which required machine harvesting, and would respond particularly to high artificial fertilizer inputs. All these characteristics foster plantation-type agriculture, which benefits multinational agribusiness corporations but not the indigenous agriculture or the masses in the Third World. Albury and Schwartz say that such research was 'geared to the aims of transnational companies seeking to create attractive investment opportunities for overseas capital'.

(c) Increased Social and Economic Control

The kind of research and development outlined above, and the techniques of legitimation used to sustain it, all lead, some think, to the increased likelihood of what Harvey (1974A) called an 'Orwellian control of citizens' lives' by the corporate state on the mass of people. The Marxist critical theory which Harvey uses conceives of the 'instrumentalist' aim of science to dominate nature as leading necessarily 'to a bureaucratic and technocratic domination of human beings' (Bottomore 1983). Thus, new technologies are often associated with the need for 'experts' in developing complex technical information and arguments. The more complex and technical they are, the less accessible are they to the ordinary citizen, and thereby it is difficult for the citizen to participate in any decision based on their use (again, see section 2.4 and 2.5.2). This is seen clearly in the nuclear debate, where representatives of the British nuclear industry have maintained that their opponents are ignorant, and that this ignorance perforce stems from their lack of sustained scientific training. Furthermore, nuclear technology is not only complex, it also demands very high investment and is potentially extremely dangerous. To safeguard the investment from saboteurs, and to safeguard the public from the damage which such people could wreak, (note the appeal to the wider public good), a cloak of official secrecy must be invoked over the nuclear industry, and security forces must be maintained which are independent from the regular police, and not publicly accountable in the same way. It has been argued that such control over the lives of the working class is another essential feature if capitalism is to thrive, and therefore that any scientific expertise which can be enlisted to extend this control will be welcomed and encouraged by capitalists. This argument has been developed by Golub and Townsend (1977) in respect of the power wielded by multinationals. Such power might conceivably be frustrated by that of individual nations if the multinationals were thought to act in a way counter to any national interest. And it could be undermined by any policy of an individual nation which might destabilise the conditions necessary for unfettered free trade and production (such as creating tariff barriers or stringent pollution controls). So anything which fostered the surrendering of some of the national power to an international body which might successfully regulate and stabilise the world economy, would be welcomed.

Golub and Townsend maintain that this is why the 'Club of Rome' welcomed and publicised selectively the findings of the *Limits to Growth* report which it had sponsored. The Club was founded on the initiative of Aurelio Peccei, an Italian business executive, with the backing of the Fiat-Olivetti Company. It developed from umbrella organisations founded in the 1960s to represent the interests of large companies (second-rank multinationals) in Europe which were at least partly owned by foreign (e.g. American) interests. They were receiving insufficient support from their own domestic governments, so that some degree of international political integration to provide a stable economic environment would have been in their interest. Golub and Townsend argue that 'the impetus for the Club of Rome's analysis and subsequent publicity campaign (which emphasised the *Limits to Growth*'s findings in favour of stability and international controls, and de-emphasised the no-growth philosophy) came from some of those firms, and individuals associated with them, who would most feel the need for international institutional economic controls'. They go on to argue that larger multinationals would also have found this in their own interest, and since they were faced by increasing instabilities and uncertainties in the economic environment, the *Limits* '"scientific" studies were commissioned as "tools of communication and control" to operate the "transitional pulley" of public opinion in order to force the governments of industrialised societies to institute a "new world moderator" which could stabilise the world economic situation and ensure a constant supply of raw materials'. In all this, the findings of an 'ahistorical objective application of the "scientific" technique of "systems analysis" were used to legitimate and support particular economic class interests, and to try to increase the control over economy and society which those interests could exert'. In support of this Golub and Townsend quote from Aurelio Peccei's book *The Chasm Ahead*:

> Nowadays all people are awed and fascinated by the new technologies they do not understand, far less dominate. In my opinion therefore, they are prepared for quite a number of years, and on condition, to recognise a new world moderator or even a new authority, set up by those who master the esoteric technologies, the Great Four [i.e. USA, EEC, USSR and Japan], even if it is a far away, supernational non-personalised and vicarious authority. ...Let us be brutal just for a moment. Once they [the Four] reach an agreement among themselves they have the power to impose it and there is no other alternative for the vast majority of the other people but to accept it.

If Golub and Townsend's thesis is to be believed, perhaps Peccei is giving us an example of what Cotgrove (1975 p73) calls 'a recent growth of quite explicitly ideological sectarian movements in science which seek to transform the world, claiming that science provides an alternative basis for authority'.

5.5 THE PARADIGM CONCEPT

Throughout this chapter we have quoted authors who see developments in science and technology as in some way 'discontinuous'. *What* has been developed, *what* scientists have chosen to do and *what* they have defined as constituting the major problems to be solved - all these have not constituted part of a smooth continuum or a linear development of science in which, impelled purely by their own curiosity, scientists have proceeded from simple to more complex problems in an unerring search for some ultimate truth. Although this evolutionary idea may be a popular image, again stemming from Bacon's man of science, the truth is more complex. Kuhn (1962) has given an insight into this complexity by proposing a model for the development of science. Although sometimes disputed, his concept of 'paradigms' in science has been widely discussed and used by historians of science in general and of particular disciplines. The paradigm model proposes that there are 'long periods in the development of scientific work during which scientists take for granted and are committed to a particular view of the world...' (Albury and Schwartz 1982). This view determines what problems are to be studied by the majority of people in a discipline. It represents a consensus among the workers in the discipline which ensures that if asked what their subject was about they would all give similar answers. This consensus is called a 'paradigm'. The paradigm 'represents the totality of the background information, the laws and theories which are taught to the aspiring scientist *as if* they were true, and which must be accepted by him if he in turn is to be accepted into the scientific community' (Richards 1983 p61). The paradigm tends to operate to determine research directions and findings (and fundings), but at any stage there may be a minority of findings which are at variance with and contradict the prevailing ethos. That minority view may be suppressed (as in Fred Hoyle's complaint that he could not gain funding for his research because it ran counter to the widely accepted exploding model of the universe) or it may gain strength because it appears to answer some of the outstanding unsolved problems more satisfactorily than the prevailing paradigm does. It may, too, derive vigour because it is taken up by the younger research workers, anxious to establish a reputation, to carve out a niche and to be noticed. Eventually the minority view may overturn that of the majority, and a new paradigm is established. The discipline has not proceeded, therefore, by slow and regular evolution: rather, it has reached its present state by a series of revolutions.

Kuhn's argument, as we have described it so far, is compatible with the notion that science is self-contained, in the sense that the impetus for new paradigm development appears to come from within - from a small group, perhaps, of scientists whose curiosity and whose findings have led them to question the consensus about what questions should be asked, and what concepts accepted. But the evidence given in this chapter suggests very strongly that such an analysis is incomplete, and that, far from being self-contained, scientific disciplines are intimately linked with the society from which the scientists are drawn, and which funds the science. Developments within a discipline - paradigm developments, that is - cannot be seen as divorced from the political, social and economic developments in society at large. The fact that the majority of publicly funded scientific R and D takes place under the umbrella of the 'defence' industries strongly suggests the truth of this. The governmental decisions to invest in nuclear energy research and to neglect 'alternative' energy research in the US and Britain seem unlikely to be independent decisions which emanated from a band of professional scientists alone.

We have said that scientific developments are linked with social and economic and political developments , and such a statement represents the use of a very broad brush. Golub and Townsend and Albury and Schwartz have been rather more specific. They have suggested that scientific developments arise in response to quite specific economic needs, and that these needs are often not generated by society as a whole - indeed, they may go *against* the broader social interest. The latter authors say, very specifically, 'The shape of particular disciplines and specialities in science and technology has been moulded by the determination of the capitalist class to mobilise all resources in their attempt to maintain ownership, power and control'. In their analysis, therefore, science and technology are greatly influenced by what happens *materially*, that is, economically, in society and that influence is exerted in the favour of one particular *class* at the expense of another. This materialist class perspective is Marxist in nature. We now need to understand it more extensively, to see how it can be relevant in understanding the political bases of the world views of environmentalism, both ecocentric and technocentric.

CHAPTER 6

THE MARXIST PERSPECTIVE ON NATURE AND ENVIRONMENTALISM

6.1 THE BASIS OF MARXISM

We have begun to see how it is that science can give us two opposing messages about man's relationship to nature. Far from being simply a source of objective facts about ourselves and the natural world, we find that science is a social activity, and that it may be ideologically based - that is, associated with sets of ideas propagated to serve the interests of particular groups in society. Having come so far, it would now be useful for us to understand more clearly this relationship between ideas and different social groups, because in dealing with environmentalism we are essentially dealing with sets of ideas, and we should see that they cannot be regarded as discrete objective entities, but must be examined in relation to the groups and the society from which they come. Marxism is a conceptual tool which helps us to carry out this examination.

Marxism takes several forms. Popularly (or unpopularly), it is regarded as a political doctrine. But it is also a conception of society and how to analyse the way in which society works. As such, it is an established intellectual discipline, and if we want to estimate the validity of the environmentalists' designs for social *change* then we first need insight into how society *works*. And though it may not be a philosophy (Edgley 1983) Marxism is a perspective which requires some knowledge of, and throws much light on, the philosophical spectrum which we introduced in the previous chapter. Hence Marxism has many implications for the man-nature relationship. We need, because of this, acquaintance with the concepts of Marxism and their relevance to some of the ideas we have already considered, such as objectivity and subjectivity, determinism and free will. Because of its characteristics, Marxism has given some writers a powerful critique of environmentalism. But to understand the critique we must first understand some of the basics of Marxism itself. To achieve this may require us to go off at a tangent to our central theme of environmentalism in this first section of Chapter 6.

Idealism versus Materialism

History is concerned with events and processes that lead to change in society, and Marx's conception of history was materialist as opposed to idealist. That is, he disagreed with those who saw social change as primarily a function of the introduction of new *ideas* into society (e.g. ideas about what constitutes progress, and about how society should function in the future). He disagreed with the view that change is a function of the growth of reason, and that it comes about through 'mental labour' - a mental struggle to create a rational society.

Rather, he thought that to understand change one needed to understand the processes by which men in society maintain themselves in existence - that is, the *material* processes of production and distribution of food, goods and services. This productive activity is a way of obtaining a means of subsistence through reacting with nature - taking the products of nature and putting them into a socially useful form. This process is *labour*, and production through labour is seen not only to change nature, but also to change man at the same time that he is engaged in it. As Sayer (1979) describes it, labour is the 'active transformation of nature for the purposes of survival' - it is the most important interaction of people and nature, and by it man transforms nature in a deliberate way, using foreknowledge of nature (unlike the way that animals transform it).

Labour is also a process of 'self creation', whereby man creates his own society through organising his work to change nature. This *material* process is the primary factor in shaping history, and the ideas and concepts by which man interprets such activity are secondary. Thus historical change and social reorganisation may (or may not) follow from a consensus about what ideas are good and what are bad, but *why* we regard some ideas as good and others as bad can be explained only by reference to our material organisation - our *mode of production*. We do not sit independently and dissociated from our material world waiting for ideas to float in and out of our consciousness, according to whether or not they appeal to our sense of rationality.

Changes in society cannot therefore come merely by appeal to people to change their ideas and values, as the ecocentrics often plead with us to do. Changes can come only after changes in the way in which we organise ourselves in economic activity - through changes in the mode of production. Such changes happen when a tension arises between the state of affairs which results from the existing social organisation, and groups which are not best served by this state of affairs. Thus the transition from feudalism to capitalism was initiated with the development of 'free' labour not tied to a feudal lord, in response to the needs of a new emerging entrepreneurial class of merchants and manufacturers. The seeds of the new system grew out of the tensions within the old, until the old was eventually replaced. This process of the negation of the present state of affairs by the seeds of its opposite is called 'the dialectic'.

If the economic organisation of society is the driving force for social change, and the concomitant, most accepted, ideas in society are closely influenced by it, we need now to examine the relationship between the two. But we should also affirm, in passing, that Marx's materialism means that to understand man it is essential to *begin* with his material conditions of production. This does *not* mean, as Schumacher (1973) takes it to mean, that Marx believed that the world consisted only of material things. On the contrary, Marx ascribed great importance to man's spiritual and non-material existence:

> In particular, Marxism must be humanist: it must give central recognition to people as distinctively subjects not objects, i.e. as beings with consciousness and values...contrary to the tendency of philosophical materialism it [Marxism] is not reductionist. It does not, that is, assert that mind, consciousness and thought are reducible to material processes and thus ultimately identical with them (Edgley 1983).

This kind of reductionism is known as crude materialism by Marxists and they will have nothing to do with it, or misinterpretations of Marx which are based on it.

The Base-Superstructure Model

So far we have suggested that everything that matters about a society is fundamentally influenced by how it maintains itself in existence. What people have to do to keep alive determines their relationship to nature and each other. So production is the central social process: the defining characteristic of being human, and it is intimately linked to social advance. When the mode of production, which is the *economic* base of society, changes, people's way of life must change.

This change comes about principally through a change in the way that people relate to one another in society. These relationships reflect considerably the economic organisation of society, therefore they are called the *relationships of production*. Thus, during the industrial revolution new forces of production (such as the steam engine) enabled men to increase their productive power, while at the same time a new industrial middle class developed, displacing the old social order dominated by a landed aristocracy. The new class consisted of *capitalists*, whose principal aim was to *accumulate capital* by producing goods and services and selling them at a price in excess of the amount of labour invested in them. The value of goods produced under capitalism - termed by Marx their *exchange value* - derived from this desire to turn commodities into money, and did not necessarily correspond to their *use value*, a value derived from how socially useful the goods were. Neither did it correspond to their *labour value*, the value derived from the amount of labour embodied in appropriating the material from nature and turning it into something useful. Labour value was a measure of how much money was required to keep the labourer and his family alive and able to produce. But there was a gap between it and the exchange value; a *surplus value*, which varied according to economic circumstances, and was a measure of how much extra work the labourer performed over and above that required to fulfil his subsistence needs. This surplus value, according to Marx, reflected the degree of *exploitation* of the worker by the capitalist. The latter was able to exploit the former because he owned the means of production, distribution and exchange. The worker owned nothing but his labour, which is, as we have discussed, his humanity - the defining characteristic of being human. He sold this - that is, he sold *himself* - in the market place. Thus the *relationship* between the worker and the capitalist - the relationship of production - was one of cash values, as determined in the labour market. The worker-to-worker and capitalist-to-capitalist relationship was also governed by this 'cash nexus', since they competed with each other economically - the one for jobs and the other for increasing ownership and control of the means of production. This relationship replaced a set of relationships in feudal society which was governed by many considerations, not resting solely on an open market for labour. There was no labour market, but in return for economic bondage between landowner and worker the former exerted a patronage over the latter which involved spiritual and welfare considerations, and was indeed in some ways benign, though oppressive.

In order to fulfil his mission of accumulating capital, the capitalist will constantly strive to increase the rate of surplus value through a variety of means (such as the division of labour) designed to secure higher productivity from the worker and/or to keep wages depressed. Also, to be successful he will strive to appropriate ever more property with his profit, enabling him to control more and more production. This is the basis for the evolution of a two-class society; *bourgeoisie and proletariat*. The former own property; the latter have only their labour to sell. Relations between the classes are governed by these economic facts, and these relationships of the economic base are translated into the institutions of society - the government, the bureaucracy, the legal systems and the political and educational institutions. These institutions and the ideas they enshrine and seek to propagate all reflect the economic (*class*) divisions of society, and function to protect and enhance the economic interests of those who own the means of production, and to curb the aspirations and expectations of those who do not. The institutions and ideas constitute the *superstructure* of society, and it is implicit in the analysis that any reform of the superstructure (e.g. in ideas and values) must be accompanied by reform of the base.

Such reform will inevitably involve (revolutionary) class conflict, since the bourgeoisie will obviously not surrender voluntarily their control of the means of production. As things stand, the ideas and institutions in the superstructure reinforce the class relationships arising out of the base - such reinforcing ideas might include beliefs in the virtues (and 'naturalness') of competition, hierarchy, survival of the fittest, the work ethic; the idea of reward in an after life (in compensation for under-reward in this); the sanctity of the family (as a unit of production/consumption and social control) and of law and order; that non-material values, and emotions, are secondary to economic 'facts of life'; that the present system represents 'freedom'. Such ideas tend to be *ideologies* - that is, they contain unchallenged assumptions which support specific (class) interests. And the extent to which they *remain* unchallenged is a function of (a) how successfully the system can maintain itself without breaking down in terms of its own economic goals (b) how successfully the media, education and other socialising influences can suppress and deflect fundamentally challenging ideas and attitudes; that is, the extent to which social consciousness can be manipulated to conform to the ideas and ideologies of the dominant class in society.

Thus,the dominant ideas in society tend to be the ideas of the dominant class. This implies that when new dominant classes emerge new dominant ideas also emerge. Conversely, one should not consider ideas in isolation from their economic base, and since the economic organisation of society has changed through history there is a need always to place ideas in their historical context, i.e. to be historically specific.

Alienation
Marx regards the relationships (of production) stemming from the (historically) specific mode of production known as capitalism as *dehumanising*. The processes of dehumanisation, together with the breakdown of social relationships and removal from nature which result from the organisation of labour under capitalism are subsumed under the concept of *alienation* - a concept which has strong if not exact parallels in ecocentrism.

The division of labour (detested by Marxists and ecocentrics alike) is the key to the concept as it is applied to production. Division of labour is one powerful way that the capitalist can increase surplus value, because it increases worker productivity, and it also deskills jobs (by splitting them into simple component parts, performed automatically). Deskilled labour is less expensive than skilled labour. Man is thus made an 'appendage of a machine', and is removed from the intellectual potentialities of labour (e.g. the skill and creativity involved in making a product from start to finish). His alienation from his own humanity (his labour) is compounded by the cash relationship between him and his employer, in which the employer does not identify subjectively with him - his thoughts, feelings and aspirations - but views him mainly in a detached way as an *object*: that is, an instrument or unit of production in the process of profit-maximisation (or, when he is a consumer, as a unit of consumption).

Furthermore, during the transition from feudalism to capitalism dehumanisation resulted from the removal of the worker from the land, so that instead of working in harmony with the cycles of nature he worked to the rhythms of the machine. He was thus alienated from *nature*. The implication of a system involving alienated labour is that the product of labour is considered to be more important than the worker himself - his value being conditioned by the cash value of that which he produces. The system also alienates the worker from himself, because his work is divorced from the rest of his life. And it alienates the worker from his fellow men, with whom he is placed in competition in the labour market. In their eyes his value is represented largely by the value of the objects which he possesses.

In these ways he is *objectified*, just as nature is objectified. He is turned into a thing, or *reified*. And the way in which he relates to his fellows, or in which nature is related to him, is via the medium of cash exchange. Marx disapproved of such cash objectification, but he recognised that some form of objectification is necessary in the evolution of society. He did not disapprove of the fashioning of objects out of nature in the production process, for this process, when it produces objects which reflect the personality of the maker and which satisfy the needs of another, produces enjoyment and makes the producer a part of the identity of the consumer and vice versa. In other words their separation into subject (consumer) and object (producer) is collapsed, and this is the essence of man's existence as a *communal* being.

Under capitalism, man's communal social essence is alienated through the formation of a *centralised state*, which projects the bourgeois ideology via control of the proletarian consciousness, and governs in the bourgeois economic interest. The state alienates people from participation in government and control of their own society, and thus people are inhibited from relating politically and socially to each other.

Liberation

Many of us are conditioned into perceiving Marxism as primarily a political doctrine, and one which is associated with apparently repressive societies. Therefore it comes as a surprise to realise the extent to which, as a mode of thought, Marxism is concerned with individual and social freedom, and liberation from the forces of alienation under capitalism. But with the establishment of common ownership of the means of production under communism and the abolition of private property comes the abolition of the private property relationships described above. This means the abolition of classes and divisions in society, and therefore of class conflict. So men reappropriate the surplus value of their labour for themselves and hold it in common. They therefore reappropriate their human essence and their existence as social and communal beings. They stop being cash objects of production and consumption in the eyes of each other and they identify with each other subjectively - for to harm the community (or nature) that one is part of by dint of ownership is to harm oneself.

Just as the division of labour was a key process in private capital accumulation, the abolition of capitalism involves overcoming the division of labour (especially between hand and brain). The purpose of labour is no longer to maximise surplus value. It involves providing the means of subsistence for the community and also fulfilment for human beings (note the very strong parallel here with Schumacher 1980). Work becomes a process of self-realisation, not just a means of keeping alive, because of the creativity involved in shaping nature to socially useful products, and because producing products useful to one's fellow beings maximises one's relationship to the community. In this way *communism* becomes a means to maximise one's *individuality* and *artistry* (note, again, the strong parallel with ecocentric ideas). Thus represented, communism differs from the crude communism which Marx disparaged, according to McLellan (1975). Crude communism is 'inspired by universal envy, negating all culture in a levelling down'. But true unalienated communist man is elevated by becoming 'total' or 'all-sided' - in other words his human faculties and attributes are able to be developed to the maximum. He is thus liberated from a capitalism which suppresses his humanity, and his latent potential is released. This is the essence of Marx's humanism - an underemphasised side of the Marxist perspective. However, it is a humanism which sees man and nature also in harmony - where nature is used for human society, but not exploited. So whereas capitalist agriculture produces deserts, communist agriculture does not, because the destruction of the land is the destruction of an asset of the whole community.

Determinism
By extension from the above, there is another important way in which Marxism becomes a theory of liberation. By revolution and the transition to communism, *man takes control of his own destiny*. He is liberated from enslavement by forces (natural and economic) which had previously been beyond his own power to control. Through taking the means of production into common ownership, man inaugurates a new era in human history, different in kind from the past because he can *'make his own history'*. What this means is that whereas in a primitive state man is strongly subject to the operation of the laws of nature, over time he is able to *control his own relationship with nature* (note that this does not necessarily imply domination, in the sense of exploitation, of nature). He becomes freer and less environmentally-determined through this historical process of self-liberation (Taylor 1982). The process probably involves a transition phase of capitalism, under which rapid technological advancement gives enhanced control over nature. But with capitalism man's capacity to control his relationship with nature can still not be fully realised, because such control can be fully exercised only by society working as a whole.

Furthermore, as a collection of objectified individuals, under capitalism men are subject to impersonal *economic laws* - the laws of the market place: of capital accumulation. They have passed into an era of being *economically determined*, and are still frustrated, in not being able to 'create their own history', that is, to control their own destiny, or future history. Nonetheless, capitalism, with its class-divided society, is still a price which may have to be paid in the transition to a liberated society. For it generates the large surpluses of production that form a necessary element in the creation of a class which is ultimately willing to take the first step towards eliminating class domination altogether. The historical process therefore represents a sequence whereby man moves from *being controlled* by natural or economic laws to having control over his society through exertion of his *will*. In this way Marxism is relevant to the philosophical spectrum (determinism-free will) which we set up earlier.

But we must beware of crude misinterpretations of Marx in this context. Thus Schmidt's (1971) presentation of Marx's concept of nature develops a model in which 'pre-bourgeois' man is dominated by nature whereas bourgeois man dominates nature. Both of these are deterministic, as we pointed out in section 5.2.2, because they promote a binary *separation* of man and nature. The pre-bourgeois-man subject is *dominated by* nature; the object. Or the bourgeois-man subject *dominates* nature, the object. Either way, man and nature are not seen in the way Marx wants us to see them - both practically and ideally as part of each other (see section 6.2).

It would be easy, too, to misinterpret the historical process in Marxism as one in which man progresses to constantly higher levels of humanity through the march of reason - becoming freed from nature, able to shape it by his *consciousness* (cf. phenomenology) and 'leaving it behind' (Burgess 1978). This would, however, correspond more to Hegel's idealism: Marx's materialist stance never saw man as becoming totally free to shape his future circumstances just as he pleased. The influence of *his own historical development* of the productive forces, including his reaction with nature, would be a constantly-present element in his society. So nature can never be 'left behind' - man's material interaction with it is a constant factor in the development of humanity. Thus men will ultimately 'make their own history', but not under circumstances chosen entirely by themselves: rather, in circumstances arising from what has gone before.

Marxism may thus offer some useful insights into the fundamental philosophical question which is so important for environmentalists - that of determinism versus free will - and we shall examine these insights further in 6.2. The question of this underlying philosophical position is made doubly important by the fact that Marxists often criticise environmentalists for being over-deterministic in their view of man both as an individual and in society - seeing him as always and inevitably controlled by 'external' forces. Such a criticism would be somewhat weakened, however, if it could be shown that Marxists *themselves* were over-deterministic. This charge is, indeed, frequently made, so we need briefly to consider its validity.

Did Marx develop a model of the historical process which holds it to be *inevitable* that man must pass through specified stages before reaching liberation - that is, is he *historically* determined? And, as an extension of this, was Marx advocating a 'scientific' theory of history in the sense of postulating inexorable historical processes, as is often claimed? Two points are relevant here:

1. McLellan's view is that Marx's model is essentially flexible. He cites correspondence which shows that Marx believed that it was possible for the stage of capitalism to be dispensed with in the transition in Russia from feudalism to socialism, and that it was wrong to suggest that there was one master key - a general historico-philosophical theory - which unlocked all history.

(a)

(b)

FIGURE 8   **The Difference Between Deterministic (a) and Dialectical (b) Relationships Between Base and Superstructure (source: Cosgrove 1979)**

2. The notion of scientific laws implies deterministic relationships of cause and effect between variables. Thus, if you change the independent variable, A (natural conditions, economic conditions, historical stages of development), then this will cause a change in the dependent variable, B (society will change, history will evolve to a new stage of development). However, Marx argued more sophisticatedly, in terms of a *dialectical* process in history, and of dialectical relationships between variables, such that a change in A influences B, but then B, in its changed state, influences A, which in its changed state further influences B, which... (see Figure 8).

Even this, however, is too simple a representation of Marx's view if it merely suggests some kind of sophisticated systems interaction of separate entities. For to Marx the two 'variables' of man and nature are *not* separate - nature cannot be regarded as external to society (see section 6.2). We must also recognise that the 'laws' of historical development, and the 'economic laws of motion' of capitalism which Marx sought to find and describe, are not to be regarded as of the same order as the laws of, say, physics. 'They are not invariant and universal laws because the societies they describe are in flux, constantly changing and developing... Far from being inevitable, these laws are socially created and can be socially abolished'. Therefore 'Marx's dialectic is a quite different mode of thought from the formal logic of positivist science, and it is symptomatic of the narrow-mindedness of the latter that its adherents often cannot conceive of different logics' (Smith and O'Keefe 1980 p32). As Williams (1983B) describes it, we are not dealing with 'categorical laws of regular determination' operating uniformly through space and time. We are concerned with 'historically specific determinations', that is 'laws' which may hold good in given circumstances - historically specific modes of production - but may not hold good in other circumstances.

Bearing these qualifications in mind, we can perhaps see that the base/superstructure model of how society operates and changes is not as deterministic as it might at first seem to be. It does not imply that man is simply the product of material conditions, nor does it postulate any simple relationship between base and superstructure. Rather, it advocates a complex, dynamic and dialectical relationship between the two (Cosgrove 1979). Kiernan (1983) develops this point, when he describes the base-superstructure concept - which Marx himself used 'only a handful of times' - as 'now felt to be too static and rigid'. A biological image', says Kiernan 'or some kind of symbiotic relationship may be more appropriate'.

The Fate of Capitalism

From Marx's own principles, detailed predictions about a future society would be baseless. For predictions are made from a viewpoint rooted in the socio-economic conditions of the time and place of the predicter. His perception is guided by a cultural filter, which itself will change. However, on the basis of the conditions which he observed, Marx did develop his theory that capitalism was destined, through its own 'inherent contradictions', to destroy itself. The theory has a strong echo in ecocentric writings such as the *Blueprint*: thus it is relevant to outline some elements of it.

First, Marx and Engels (in the *Communist Manifesto*) described how the bourgeoisie (capitalists) revolutionised production. They did this by creating the 'cash nexus' - putting an 'end to all feudal, patriarchal, idyllic relations', and 'pitilessly' tearing asunder 'the motley feudal ties' binding man to his 'natural superiors'. Thus 'freed' to sell his labour on the open market, the worker's personal worth was resolved into exchange value, while he was also forced to move to towns that acted as a focus for the labour market. However the workers thereby became the 'weapons that bring death' to the bourgeoisie, because the latter did away 'with the scattered state of the population, of the means of production, and of property', agglomerating people and centralising the means of production. In this process the proletariat were gathered together in great factories, in which form they became privates of the 'industrial army'. They then began to form combinations (trades' unions) and political parties, which could challenge the attempts of the bourgeoisie to hold down wages and conditions. Thus the bourgeoisie's attempts to advance industry had led to the creation of an *organised* proletariat, with a consequent power that would have been denied them in their 'scattered state'. As such, they were to become the means of the downfall of the bourgeoisie, so through this contradiction the latter would become 'their own gravediggers'. The strength of proletariat organisation was increased by another contradictory element, whereby because of the bourgeoisie's constant attempts to monopolise ownership of the means of production such ownership became centralised into fewer and fewer hands. Thus the numbers of bourgeois decreased as those who lost out in the monopolistic competition (the 'petty' bourgeoisie) fell into the swelling ranks of the proletariat. As they descended they brought with them a theoretical understanding of what was going on - supplying the proletariat with 'fresh elements of enlightenment and progress', and 'weapons for fighting the bourgeoisie'.

All this took place because of the inevitable desire of capitalists, by definition, to increase the surplus value of production by making the worker more productive. As the latter is more productive through centralised large-scale means, so is he also more productive through the division of labour. As described above, such a division leads to deskilling, so that 'In proportion, therefore, as the repulsiveness of the work increases, the wage decreases'. Repulsive work is not done for any intrinsic satisfaction but instead the proletariat's attention is fixed on the cash value of the work - increasing the dissatisfaction which the trades' unions articulate. The labour-dividing process is aided by the use of machines, and is obviously another inherent feature of capitalism, whereby unemployment is inexorably associated with the capitalists' desire to maximise productivity and thereby surplus value. Finally, in their desire to maximise capital accumulation, the bourgeoisie was rightly seen to need 'a constantly expanding market for its products'. This need led to the destruction of national boundaries and characteristics and the global establishment of capitalism. 'In place of the old wants, satisfied by the productions of the country, we find new wants, requiring for their satisfaction the products of distant lands and climes. In place of the old local and national seclusion and self-sufficiency, we have intercourse in every direction, universal interdependence of nations'.

Many environmentalists as well as Marxists would say that this market enlargement, through creating 'new wants' and opening up new territories, has led to the contradiction of overproduction of material goods existing alongside unmet social needs. Overproduction results from an inherent contradiction of capitalism because labourers cannot consume as much as they produce, since they always produce amounts with an exchange value in excess of what they are paid for their work. There will be constant economic crises, therefore, which are resolvable only through restoring an equivalence between sales and purchasing power. In this process, capitalists go out of business.

In the modern context, capitalism has produced a Third World proletariat, placed in a subordinate relationship to predominantly Western capitalists. But it has also brought a depletion of the natural resources which are necessary for the maintenance of the capital accumulation process. And as this depletion becomes increasingly seen as resulting from particular economic relationships, perhaps the political consciousness of the Third World proletariat will be raised in a way analogous to that of which Marx and Engels wrote.

The Role of the State, and the Idea of Pluralism

In examining the relationship between economic base and institutional infrastructure, Marx was impressed by how much the laws of a nation reflected the interests of those in power (the basis for political power being ownership and control of the means of production). This has implications for those environmentalists who believe in the power of ideas, developed by debate within a pluralist democracy, to change society. Marx opposed Hegel's view that human consciousness manifests itself *objectively* in man's judicial, social and political institutions, organised together as the state. In this view the state would harmonise the conflicting interests of groups with each other and with 'universal reason'. But Marx said 'Just as religion does not create man, but man creates religion, so the constitution does not create the people but the people the constitution'. And as the people are divided into conflicting classes, so the bureaucracy which upholds the constitution reflects this division. It does not perform a mediating function between different social groups, acting as a 'universal class' in the majority interest. Instead, the bureaucracy encourages the political divisions which are essential to its own existence (McLellan 1975). Thus the United States Constitution does not represent a set of eternal truths. It must be understood in the historical context of the demands by new commercial groups for the end of feudal restrictions on their activities and for 'free' competition in economic affairs.

Thus to Marx, an opponent of the state, the state represented the 'projection of the alienated essence of man', giving an illusory sense of community. It was a screen for the class struggle which derived from the mode of production. 'Political power, properly so called, is merely the organised power of one class for oppressing another... The executive of the modern state is but a committee for managing the affairs of the whole bourgeoisie'. And the state, as manifest in the bureaucracy, the judiciary and the standing army, functions to keep the majority of producers beneath the yoke of the minority of exploiters. With revolution, and the abolition of the division of labour, should come the abolition of the distinction between those who govern and those who are governed, relegating the bureaucracy's role to a purely administrative one while men regain their communal essence by means of their ability to shape society through a true participative democracy.

Clearly if Marx was correct in this analysis of the role of the state and the bureaucracy, the extent to which any serious and fundamental challenge to the prevailing order can be mounted via the institutions of the state (e.g. by challenging the growth assumptions behind major environmental projects in public inquiries or through Parliamentary lobbying) must be severely limited (see section 6.5).

6.2 HOW MARXISTS SEE NATURE
6.2.1 The problems of Determinism and Free Will
Now that we have expressed attitudes to nature in terms of two basic and opposed philosophical 'extremes' of determinism and free will (5.2.2), it is wise to begin this outline of the Marxist view of nature by reference to that philosophical spectrum. We see that Marxism highlights the difficulties of adopting either extreme; and that in one sense it adopts a

midway position within the spectrum, but that in another sense it transcends it altogether, by dismissing it as irrelevant.

Burgess (1978) describes the Marxist concept of nature as midway between classical materialism and classical idealism, being a synthesis of the two traditions. That means that it combines elements of an approach which stresses the effect of human (economic) activity upon nature - and vice versa - in a deterministic way, with idealist elements which see nature as essentially a manifestation of human consciousness - human will - and therefore susceptible to being changed by changing ideas. However, it does this in such a way that the consideration of environmental attitudes and behaviour is not a *speculative* one, steeped in abstract philosphies. Rather, it seeks to establish the connection between attitudes towards nature and the day-to-day *practice* of man's economic activity, seen throughout history. As Quaini (1982) expresses it, determinism and free will are two paradoxical viewpoints, and we should escape from the paradox via historical materialism. This approach shows that the sensuous world around us (nature) is not a fixed thing for all eternity, but varies in relation to man's industry and the state of society - what he *does*, in practice. So historical materialism is a kind of *anti-philosophy*, resolving speculative problems of the man-nature relationship into real, empirical ones. Nature in the abstract is anyway of no concern to man: nature has interest only as a sphere of human activity (to see the force of this we might try to define 'nature' without referring to 'man', and we will see that this is virtually impossible). So Marxist method turns the 'philosophy' of nature into a 'history' of nature and mankind - grounded in what can actually be *observed* about the man-nature relationship through time. In this way it does not lose sight of the 'historicity of nature' (how the concept of nature is related to human activity) or the 'naturalness of history' (how human activity is related to natural conditions).

Indeed, Marxism holds that man and nature are not separate and are not governed by a subject-object relationship. Man is part of nature and vice versa. However the unity of the two is not substantially of a mystical phenomenological kind (where nature is merely an extension of human consciousness); rather, it comes about via the process of *labour*, as will be explained shortly. In a material sense, there is no nature anywhere which 'precedes human history', i.e. does not bear the imprint of, and is not substantially moulded by, human society. To say this 'does not imply that every atom of some tree, mountain or desert is humanly created, any more than every atom of the Empire State Building is humanly created...' but 'It does mean that the human activity is responsible to a greater or lesser extent for the form of matter; the size and shape of the buildings, the hybrid or location of the tree, the physiognomy of the mountain, the spatial extent of the desert'. And in terms of ideas 'We cannot know nature as external; we can know it only by entering into a relation with it' (Smith and O'Keefe 1980).

The Marxist objection to pure determinism takes two forms. First, there is an aversion to explanations of historical development which are derived from *extra*-social and *extra*-historical factors, on the grounds that they are politically reactionary because they encourage the *status quo*. For if human options are substantially a function of climate, race, the Darwinian struggle, or other 'natural' laws and forces, then opportunities to change society in the future are strictly limited (see sections 5.3, 6.4 and 7.3). This feeling seems to have underlain the attacks which were made in the USSR in the 1930s on some Soviet geographers who committed the sin of 'geographical deviation' (environmental determinism as practised by Western geographers), when The Party under Stalin formulated its ideas about the forces of social development. Determinism was denounced as a 'scientific weapon of the bourgeoisie' and the foundation of racism and justifier of war (Matley 1966). Even in the 1960s and early '70s when Anuchin and Saushkin argued for recognition of the importance of environmental factors in social development, and for forms of unified study of society and nature, they were attacked on the grounds that it was 'not nature that shapes human society but the productive relations among people' (Matley 1982).

Carried to extremes, this Soviet attitude resulted in the disastrous Stalinist dictum that the geographical environment affected only the speed of social development; it was not a determining factor, and man was the master of nature. When ecocentrics like Fry (1975) and Bookchin (1980) attack Marxism, this is what they are attacking - apparently unaware that Marxists themselves also hit out at the 'idiocies of the Stalinist approach' (Burgess 1978). These 'idiocies' are founded on a crude materialistic interpretation of the relationship between the economic 'base' and the superstructure of society, leading to the idea that *everything* flows deterministically from economic organisation - including nature. This is the 'classical bourgeois misinterpretation' of Marx, committed by the USSR (which some people erroneously call a socialist state). It sees dialectical materialism as meaning that the universe can be explained simply in terms of matter in motion. In this type of explanation both man, and productive forces like nature, are viewed mechanically, as things or objects related in a simple deterministic way. Herein lies the second objection to determinism, as 'vulgar materialism'.

Sayer (1979) expresses this objection by distinguishing between human and other life forms. Humans, he says, interact *socially*, and not just mechanically. This means that their actions are not merely behavioural (semi-automatic), but they have meaning. Arguing, voting, getting married: these are actions 'whose nature depends on the existence of certain intersubjective meanings' which are attached to the actions. The problem with crude, mechanical explanations of society is that they fail to recognise the importance of the *meaning* of actions (human intentionality). This leads to a treatment of people and nature as objects - reducing them to mere data, and producing purely technical evaluations of the relationships between them, and with nature. All behaviour is reduced to positivistic causes and effects (deterministic relationships) and the *reasons* for the behaviour and whether these are acceptable (values) are ignored. Marxism, however, distances itself from these pitfalls by distinguishing between reflexive behaviour (e.g. of animals) and purposive action (of man). In so doing it avoids the worst excesses of positivist determinism.

However it also divorces itself from the idealism of the other philosophical extreme. For the meanings of the ideas behind our actions are said to *derive from the society we live in,* and are not wholly reducible to the opinions of individuals (see the description of existentialism and phenomenology in section 5.2). These meanings 'are not free emanations of the thought of isolated individuals, nor are they the simple responses of isolated individuals to nature'. They are connected to a specific structure of social relations which have a material basis (Sayer 1979). The problem with idealist approaches like phenomenology is that reality (nature) is seen as moulded by man's consciousness. So man becomes internally self-sufficient. And the 'whole effort of man shaping himself and his world' is seen 'as being dictated by consciousness, and an individualised consciousness at that. History becomes a manifestation of the act of thinking and nature becomes a quality of that thinking' (Burgess 1978). So man is removed from his social and natural context, yet it is via *society,* through particular forms of production, that his relationship to nature is formulated. Furthermore, phenomenological approaches limit our investigations of nature to descriptive appearances, and do not penetrate to the objective reality which, Marx maintained with irony, *is* there:

> Once upon a time a valiant fellow had the idea that men were drowned in water only because they were possessed with the idea of gravity. If they were to knock this notion out of their heads, say by stating it to be a superstition, a religious concept, they would be sublimely proof against any danger from water. His whole life long he fought against the illusion of gravity, of whose harmful results all statistics brought him new and manifold evidence (Marx and Engels: *The German Ideology,* quoted in Ley 1978).

In the Marxist conception of nature, the notion of the transformation of reality through consciousness is, in fact, retained but it is subjected to what Burgess calls 'a set of materialist premises'. Thus reality is perceived by the mind, but it is not *merely an expression* of the mind; 'there exists a natural substratum which is a necessary precondition for the activity of consciousness'. In other words, our perceptions tell us what reality is like, but these perceptions themselves are not born of isolated abstract ideas: they relate to the social (class) relationships of the people who have the ideas, which in turn relate to the process of production. In this way, according to Quaini, Marx describes an extra-human reality which is both independent of man and capable of being mediated by him. Hence Marxism avoids both 'spiritualistic monism' (everything as ultimately the expression of the spirit, or ideas) and 'naturalistic monism' (everything is ultimately determined by nature). The essence of Marx's man-nature 'dialectic' is that a particular behaviour of man towards nature is conditioned by the form of society, but that also the form of society is conditioned by man's behaviour towards nature. The crucial and most fundamental expression of this man-nature interaction is, according to Sayer, through labour.

6.2.2 The Man-Nature Dialectic

Labour is the means whereby man converts nature into forms useful to him. In the main nature does not offer ready-made subsistence to man, neither does man take direct possession of nature's resources. He has to transform them. This process of transformation is a *social* one - it is done with other people who are organised in a particular way. So the process of changing nature involves forms of social organisation, which give rise to particular relationships among people (capitalist forms of organisation lead to relationships expressed in cash terms). Thus, through shaping nature, men shape their own society and their relations with their fellows. This happens all the time: it is, as Burgess says, 'a constant total interaction of subject and object in the historical process'. In fact there is *no* subject and object in the classical deterministic scientific sense. Both 'sides' act and react on one another in an organic sense: man sets in motion

> arms and legs, head and hands, the natural forces of his body, in order to appropriate nature's productions in a form adapted to his own wants. *By thus acting on the external world and changing it, he at the same time changes his own nature.* He develops his slumbering powers and compels them to act in obedience to his sway (Marx, *Capital*, 1965 ed. p173, quoted in Quaini 1982).

The emphasis added here sums up the dialectic. The Russian geographer Saushkin illustrated it in 1973 in a way which, says Matley (1982), accords with the current official Soviet view. The transformation of nature has allowed an expanded reproduction of productive forces. This in turn has enabled society to exist with higher population levels at higher levels of development than would be possible were the environment untransformed ('natural'). In fact the effects on society of an untransformed (e.g. without agriculture) environment are quite limited, because it contains only natural characteristics. Transformed nature (e.g. agriculturalised land, minerals after mining) has a far more fundamental effect on society.

However, this example of self-transformation via transforming nature is set in purely material terms. Burgess also pays attention to the 'existential quality in the Marxist system of thought' when the man-nature relationship is mediated through labour. Man works on nature, he says, to produce a world of objects. But this work differs from that of animals in gaining subsistence, for the objects which humans produce are, with time, not merely to satisfy basic physical needs. Through gradual interaction with nature man changes his character from one who is merely a satisfier of basic needs to someone who 'sees the beauty' in the objects he produces. This constant interaction makes him more *human* over time. 'Thus the dialectic is made clear: as the world is increasingly more humanised so too are the senses humanly developed as a social process' (Burgess 1978 p2). One might add, as Sayer points out, that human, as distinct from animal, labour is done with meaning and intentionality. Therefore man conceives in advance what he wants to produce; how he wants to transform nature. In so doing he requires foreknowledge of how nature 'works' - its mechanisms and laws. This knowledge enables him to produce for more sophisiticated needs. In the process of knowing nature to transform it, man transforms himself to a higher intellectual plane.

As he gets away from the process of merely satisfying basic needs, he creates new needs (e.g. for mobility or leisure), and possibilities for their satisfaction, as we have pointed out. Needs, apart from the most basic, are not, then, objectively definable for all time. They are a function of stages of historical development and of man's consciousness at any given stage - that is of his perception of what he needs and of what, therefore, is a resource to satisfy the need. (For example, the need for motor fuel and for plastics from oil did not exist until it became realised that oil was not just a black, sticky and objectionable nuisance seeping out of the ground, but could be *used* in particular ways.) Questions, then, of how we define a 'need', and a 'natural resource' are time-based: they are relative to historical stages of development; they are historically specific and related to particular modes of production. And it is *labour*, with its resultant transformation of nature, which makes the development of society historical in this way and not just a matter of (preordained) biological evolution. Social development, with its changes in the relationships of production consequent upon changes in ownership and control of the means of production, is intimately bound up with the social action of transforming nature in organised labour; acting upon socially-derived knowledge of nature's laws, and upon socially-based perceptions of what is needed from nature and what nature can offer.

To repeat, the time element cannot be left out of any Marxist consideration of nature and man's relationship to it. Different forms of perception and modification of nature correspond to specific historical stages of human development. This, then, is what is meant by the term 'historicity of nature', and we see today a 'nature' that almost entirely bears testimony to the intimacy of the man-nature relationship over time: a historically-produced nature, because virtually none of the world's landscapes or ecosystems are uninfluenced by man.

In the dialectic between nature and man, neither, ultimately, becomes the subject or the object. They exist in an organic intimacy of constant interaction. Marxists see social relationships in the same light, and they object to capitalist social organisation, which produces subject-object relationships between people. Burgess encapsulates the objection as the problem of the 'I' and the 'we'. He says that ever since Descartes, 'I' has become the 'primary datum' in Western society, rather than 'we'. That is, questions of human relationships become translated by people into the problem of 'the other', of 'other people', rather than problems to which the subject, 'I', contribute. 'And', says Burgess, 'these other men become reified as things that I observe and understand, just as I would, say, a housebrick'. But:

for Marxism on the other hand, men are conceived of not as reified objects that are to be observed passively, but as subjects - men with whom 'I' act collectively. Man is by definition a species-being, he is inherently social, and this social nature is expressed in his interactions with nature and his fellow men (Burgess 1978 p9).

However the time perspective is again important. For this subjective Marxian relationship between men - and between man and nature - is not going to be possible except in the classless socialist society (bereft of the cash nexus) towards which we are moving.

6.2.3 Pre-Capitalist and Capitalist Man-Nature Relationships

To summarise, we have to realise that in the Marxist conception of nature specific states of human consciousness about the world correspond to specific social relationships which stem from production - relationships which are summed up in the concept of class. Since class relationships change through history, man's relationships to nature and his fellows are intrinsically social and *historical*. This means, says Burgess, that to understand consensual views about nature we would do better not to fix on individual perceptions (e.g. on the ideas of 'great men'), but to examine the social and historical bases of the consensus - the way it reflects and reflected class structure and class interests arising from how production (transformation of nature) was organised. In other words, we should look at the historically specific and always social concepts of nature and how they are and were 'related to the social totality which emits them' (Burgess 1978).

When we do this we see that in any given historical period the interaction with and perception of nature is 'locked into a determinate structure of social relations' (Sayer 1979) produced by labour. As we discuss below, when we pass from feudalism to capitalism nature becomes perceived as something more than just of social use value. Through labour organised to produce exchange rather than use value, it now becomes a *commodity*, and seen in terms of cash, just as all social relationships are so viewed. So we see that immediately we take on this historical perspective we can avoid the pitfalls of universalisation - believing that the social relations and the interaction with nature that are evident at present have always obtained and always will - that they have an 'eternal necessity'. This kind of universalisation would mean that we would tend to see historical and social conditions as 'natural' and subject to an evolution which is controlled less by human will than by factors such as *laws* of 'human nature' or 'natural events and conditions'. But Marx points out that it is only when the labour invested in a product is regarded from the viewpoint of the product's *exchange* rather than *use* value, that labour (and nature) become estranged/alienated/objectivised - a function of impersonal 'laws' of economics which appear universal but in reality are specific to capitalism. As Quaini (1982 p39) puts it, nature spills over into history. Capitalism on the one hand technologically dominates nature, but on the other creates an apparently 'natural' society 'which sets itself against people and rules them more than a natural nature would have dominated pre-capitalist societies'.

Quaini interprets Marx's (1981 p471) view of 'forms which precede capitalist production' as being dominated by use rather than exchange values. This led to a very different conception of nature from that obtaining under capitalism. But the 'freeing of labour' from slavery and serfdom under capitalism meant that labour was exchanged for money. This involved separating labour from the 'means of labour and the material of labour... above all, the release of the worker from the soil as his natural workshop'.

However in pre-capitalist forms man was linked to the land in such a way that it 'presupposed his own activity' - nature was 'just like his skin, his sense organs' (Marx 1981 p490). He did not stand in this relationship as an individual but as a member of a community: his appropriation of nature was mediated through the commune. In these agrarian societies the relationship to nature was more dominant than historical/social factors. Thus, on a feudal farm climate substantially helped to determine the volume of production in a year, whereas through technology a capitalist farmer is freer from the climatic influence, and he is more preoccupied with the pattern of prices in determining what will constitute his 'marketable surplus'.

In the mode of agrarian production which capitalism eclipsed, and which had small free land owners and communal ownership, men, says Quaini, related to nature as their property. They were masters of the conditions of their reality, having an existence independent of simply being labour. They related to each other as co-proprietors; as members of a community where the main aim of the work was not the creation of exchange value but the sustenance of the community. So production was for man, rather than man 'being for production' - i.e. with the purpose of generating surplus value for profit. Quaini's (p57) view of capitalist history is 'the story of the separation of the producer or worker from his means of production...or in other words...man's expropriation from nature and from the primitive natural community'. This progressive dissociation of man from his environment took place as land was transferred from use to commodity value. *Both man and nature became commodities, acquiring abstract alienated existences.* Social and nature-man relationships became not personal, but as between *things*. This was the negative and contradictory side of the scientific conquest of nature, which under capitalism became transposed from its original Baconian 'purpose' of improving the lot of all mankind to a specific one of advancing the interests of the bourgeoisie. As Smith and O'Keefe put it, 'Under capitalism, the relationship with nature is a use-value relation only in the most subordinate sense. Before anything else it is an exchange-value relation'. Thus 'expansion', 'growth', 'progress', and 'development', which have come increasingly to mean the accelerating accumulation of greater quantitites of capital, also and *therefore* have come to mean the affixing of exchange values to realms of life - including nature, art and religion - which had previously been '"sacred", that is, had survived independently in the realm of use values'. But now religion, art and nature are becoming commodities as much as scientific knowledge already is. The mark of progress is to produce and consume more commodities, and 'Whether we are talking about the laborious conversion of iron ore into steel and eventually into automobiles or the professional packaging of Yosemite National Park, nature is produced... The form of all nature has been altered by human activity, and today this production is accomplished not for the fulfilment of needs in general but for the fulfilment of one particular 'need': profit' (Smith and O'Keefe 1980 p35).

The man-nature separation and alienation has not therefore been one merely of the realm of ideas - not found simply in abstract concepts like the subject-object dualism (Chapter 2). Capitalism's development from feudalism was accompanied by the real growth of the modern form of industrialisation which had an obvious visible manifestation in the large-scale manufacturing city (but is also, in the late 20th century, becoming more and more a feature of rural landscapes). This was industrialisation based on capital accumulation - fundamental to which process was 'the expropriation of the rural population and its expulsion from the land, an enormous phenomenon of the separation of man from nature, of the producer from his workplace' (Quaini 1982 p103). The expelled rural population became an urban proletariat and city and country grew further apart - a move founded on the division of labour brought about by commodity exchange. The division was a spatial one, concentrating labour into factories, leading to massive urban growth, and an upsetting of rural-urban 'balance'. In turn this led to pressure on the land to produce more. So:

> By the incorporation of the countryside into the system of production of values and goods for trading, capital gradually subordinated the whole land to itself and to its urban centres of power assimilation (Quaini p109).

We must stress that in this analysis the man-nature separation and the growth of social alienation which are so decried by ecocentrics took place under a specific (capitalist) organisation of production, i.e. it was produced by industrialisation under capitalism, not industrialisation *per se*. So it is misguided to rail against the latter as the fount of all ecological evil, as some ecosocialists (see 7.2.6) and many romantic ecocentrics do. This is a misguided response because it is *ahistorical* (see section 6.4).

Marxists like Quaini are scathing about romantics - and also 'idealistic socialists' like Levi-Strauss - whose representations of an idyllic primitive society, organically bound with, and not exploitative of, nature (cf. Goldsmith 1978) are dismissed as 'wild talk'. 'Populist reactionary ecology' tries to turn the clock back, with idealist, simple-minded philosophical mystification. To speak of introducing a balanced man-nature relationship into a *capitalist* society is a 'wild hope'.

In reality, modern science has removed any validity from the childlike perception of nature inherent in romanticism. Thus to make their own stance realistic and constructive, modern ecocentrics must, in the Marxist view, follow Marx in rejecting *both* the despoliation of nature under capitalism, *and* the deification of nature.

6.3 HOW MARXISTS SEE SPECIFIC ENVIRONMENTAL PROBLEMS

6.3.1 'Overpopulation'

The theoretical perspectives described above demand that current problems of population numbers in relation to resource availability must be seen in a *historical* perspective; that is, in relation to specific modes of production. The conclusion of such an examination is that these problems, as well as that of environmental abuse, can best be solved within a socialist mode of production.

Marxists say that only animals and plants are bound by an 'abstract' - that is, an objective, unchanging, non-historical law of population. But where human societies have little power to change their environment population size *is* an important limiting factor on material wellbeing because of physical limits on resources. However, where societies can change and manage their environments the population size which can be sustained is determined more by social relations. Thus we can talk not of one universal law of population, but of a law of population which is specific to a given mode of production, like capitalism, and is bound up with the social relations of production.

Marxists also ask what a term like 'overpopulation' means - 'overpopulation' in relation to *what*? And how do we know when there is 'overpopulation'? We may say that overpopulation is evidenced by the existence of groups who do not have enough to eat (they are the surplus people), which is what resource 'scarcity' implied in Malthus' time (Perelman 1979), but it does not really follow that this starvation is produced by 'natural shortages' i.e. an *absolute* inability of the earth to produce more food. Rather, the 'surplus' population may not be able to buy food simply as a result of the inability (or unwillingness) of an economic system to create enough jobs or to pay enough to those who do work.

In fact, as Marx showed, it is fundamental to the nature of capitalism that wages must be kept as low as possible, and that to do this there must be a pool of unemployed labour (section 6.1). To maximise capital accumulation, that is surplus value, wages must be kept depressed in relation to the exchange value of products, even when the latter may be increasing. One way to do this is to have an unemployed pool of people who will be ready to step in and do poorly-paid jobs if those already in them threaten to strike for more wages. Competition for jobs keeps wages low. (This is what has happened in Britain in the early 1980s, when the unemployed pool of 3 million has led to reduced wage demands, and even some willingness to take pay cuts - see section 8.2.) The production of the pool, the *industrial reserve army*, is an inevitable outcome, in any case, of the constant tendency for capital to be substituted for labour. So the reserve army of the unemployed, produced by mechanisation in order to maximise capital accumulation, permits the expansion of the surplus value produced by labour, and also furnishes a pool of labour to draw on in times of boom. The members of the 'army', however, *appear* normally to be superfluous to requirements, even though they fulfil this vital function for capitalism.

They cannot buy enough, so *for them* resources *are* scarce. And by keeping the wages of others depressed, this relative surplus population serves to give an even wider appearance of Malthusian 'starving overbreeding masses'. But poverty is not said by Malthus to result from this politico-economic process. Instead it is put down to 'natural' population laws, which are unavoidable except through the efforts of the working classes to restrain their own passions (see 5.3.2). However:

> The production of a relative surplus population and the industrial reserve army are seen in Marx's work as historically specific - internal to the capitalist mode of production. On the basis of his analysis we can predict the occurrence of poverty no matter what is the rate of population change (Harvey 1974B).

Harvey goes on to illustrate the value of an historical perspective on this problem. He examines a particular sentence containing some words which often have a ring of the absolute. The sentence is: '*Overpopulation arises because of a scarcity of resources available for meeting the subsistence needs of the mass of the population*', and it resembles the typical pronouncements of neo-Malthusians. Harvey focusses particularly on the words 'subsistence', 'resources' and 'scarcity'. The first, he says, has in fact a relative meaning, which is socially and culturally determined. 'Needs' can be created, implying that they are not purely biological, and thus that what constitutes 'subsistence' needs is internal to a particular society, with its particular mode of production. 'Subsistence needs' therefore change over time. Similarly, 'resources' are things in nature which can be 'useful' to man. Thus they can be defined only with respect to a society's stage of technical development - i.e. its *ability to use* nature's materials. As Spoehr (1967) says 'It is doubtful that many other societies, most of which are less involved with technological development, think about natural resources in the same way as we do'. And 'scarcity' is not inherent in nature; its definition is 'inextricably social and cultural in origin' because it can be assessed only in respect of what a society wants to attain in the first place (e.g. mere biological subsistence or high economic living standards). The original sentence can thus be rewritten as:

> There are too many people in the world because the particular ends we have in view (together with the forms of social organisation we have) and the materials available in nature that we have the will and the way to use, are not sufficient to provide us with those things to which we are accustomed (Harvey 1974B p236).

This redraft makes population/resource problems no longer insoluble by any human means other than by altering our population numbers. For we could change the ends we have in mind, and social organisation ('scarcity'), or we could change our technical and cultural appraisals of nature ('resources'), or we could change our views concerning the things to which we are accustomed ('subsistence'). For Harvey, the problem about focussing discussions on whether or not we change population numbers lies in their unacceptable ideological content. 'Whenever a theory of overpopulation seizes hold in a society dominated by an elite, then the non-elite invariably experience some form of political, economic and social repression' (see section 7.3.3).

6.3.2 Resource Depletion and Pollution

So one facet of the Marxist approach to questions of resource availability and subsistence is to look hard at these terms themselves and set them in their economic and social context, rather than investing them with notions of universality. One can readily see how, in capitalism, the idea of a need (and hence resource availability to fill it) is highly contingent on social relations of production. 'Need' is not usually expressed in terms of what would be socially useful to *all* of the people, but in terms of the aggregate of individual 'demands', or wants, expressed by *those with appropriate purchasing power*. This way of assessing 'need' is peculiar to capitalism, however. Hence it is not easy to determine what the *social* needs of a capitalist society might be at any given time. 'The system, apparently, cannot be judged because it does not produce the criteria by which to judge it' (Emerson 1976A), except, that is, the criterion of ability to pay.

However, many of the 'needs' which *are* satisfied by the capitalist market mechanism may be deemed frivolous - so much so that much money has to be spent to persuade people that they have such needs (Galbraith 1958). Emerson (1976B) analyses the main components of the UK GNP, and decides that most of them constitute rather debatable 'needs', such as private cars, cigarettes, tobacco and alcohol, radio, electrical and other durables, cosmetics and toiletries, roads, 'defence', overconsumption of fuel and light, and of food. One-quarter of food produced in Britain is dumped after being processed or cooked, while a lot more goes to supply EEC 'mountains' - surpluses (held in store in order to maintain artificially high price levels) which may ultimately be destroyed. A similar situation obtains in North America (Sandbach 1978B). This is one aspect of the process of constant production expansion inherent in capitalism, while waste creation and built-in obsolescence to create enlarging demand is another (Packard 1963).

Marxist analysis of capitalism has encouraged us to see current resource problems, then, as specific to capitalism. And even accepting that resources were to be in some way absolute - that there was an upper limit on their availability - Marxism would still encourage us to see the problem of their depletion in relation to specific modes of production. The capitalist mode is characterised by a growth 'treadmill' which is produced by a contradiction. The contradiction is that to maximise surplus value, the capitalist must see that the exchange value of a product is above what he pays for the labour invested in it (see section 6.1). However, he also needs to sell all his products, so it is necessary for the labour force to have enough purchasing power to constitute a robust market. He will nonetheless be tending all the time to *reduce* the aggregate purchasing power by his desire to substitute capital for labour, producing unemployment. In order to cope with this - to

169

provide enough work to have a workforce that can buy his products *and* to create expanding surplus value (it must expand in order to allow accumulation of capital to permit reinvestment) *production must constantly expand*. This expanded production must be absorbed by consumers. More consumers must be brought into the market, by spatial expansion, and/or higher incomes must be paid to existing workers. But this latter course tends to reduce profit margins, so necessitating further expansion and merely continuing the process whereby the workforce's purchasing power must be increased in order to buy the increased output (Schnaiberg 1980 pp228-9). Of course complicating factors like state intervention and the existence of unproductive 'consumers' like the defence sector enter into the picture, but in general we can say that if, like some ecocentrics, we wish actually to abolish *growth* based on ever-increasing consumerism - the consumption of commodities - we have to abolish *capitalism* itself. The only way to achieve a stable-state economy in the modern world would be to equate the exchange value and social use value of products, and to narrow the gap between these values and what we pay for the labour they represent.

But, whereas some ecocentrics might disapprove of growth *per se*, Marxists do not. In the first place they see that the 'full development of modern productive forces' is necessary before socialism can be established. The division into classes is regarded as an inevitable feature of societies which have insufficient production, so growth towards a sufficiency is needed before people will begin to think of their future society rather than just immediate need satisfaction. Secondly, it is *commodity* production and consumerism which are seen to create problems, rather than production for democratically-determined social needs. The former lead to resource squandering, and this is a capitalist feature even though it also happens in planned economies, which have become hooked on consumerism (Barratt Brown 1976). But growth to provide for social needs is a desirable feature.

However, this kind of growth cannot satisfactorily take place merely through the operations of the free market. For the latter has simply produced Western-style consumer abundance existing alongside unmet social need in the developed countries and alongside Third World starvation.

Quaini (1982) reminds us of how Marx and Engels (in the Communist Manifesto, for example - see section 6.1) described the perpetual search for new markets in order to try to solve the contradictions described above. They said that through this search the entire earth would be uniformly 'civilised' via the process of need-creation. In it, nature would become objectified - purely a matter of utility - and conquered. And old national identities and barriers would be destroyed in the creation of this universal society. This was, then, the state of 'permanent revolution' stemming from capitalism's search for new forms of production to create new 'needs' and markets, leading to the alienation of people from each other and from nature.

But this alienation, says Quaini, is another contradiction. For it becomes, in fact, a 'barrier to capital'. It produces in the end the destruction of the natural resources of which capitalism has continuing and growing need. Quaini illustrates this by reference to Marx's discussion in *Capital* of the resource of soil fertility - a discussion which seemed remarkably to pre-empt contemporary concern over the effects of modern ('green revolution') farming techniques on the soil. Marx noted that the development of capitalism collected the population into great centres and

> disturbs the circulation of matter between men and the soil i.e. prevents the return to the soil of its elements consumed by man in the form of food and clothing; it therefore violates the conditions necessary to lasting fertility of the soil...all progress in capitalistic agriculture is a progress in the art, not only of robbing the labourer, but of robbing the soil; all progress in increasing the fertility of the soil for a given time, is a progress towards ruining the lasting sources of that fertility... (Marx 1959 pp505-7).

Marx concluded that capitalist agriculture was not rational - it mindlessly left 'deserts' behind it as land was destroyed by the profit-seeking bourgeoisie. This image of a desert is not out of place in modern discussions of what capitalist farming ('agribusiness') is doing to the soil. Quite apart from the literal desertification of savannah regions of the Third World which has resulted from the importation there of Westernised agriculture, there is great concern in Britain over the prospect of declining middle- and long-term soil fertility. This is seen to be a danger stemming from the short-term search for increased 'efficiency' - that is profit maximisation in relation solely to food prices, rather than social or environmental need. With its high inputs of artificial fertilisers, chemical pesticides and herbicides, and machinery, and with its intensive methods, modern farming produces eutrophication of rivers, water and land pollution with antibiotics used in animal rearing, deterioration in soil drainage and structure through (over)ploughing in wet conditions, a loss of porosity through the destruction of worms by copper oxychloride pesticides, structural weakness through the substitution of artificial for organic fertilisers, soil erosion following on hedgerow removal for larger more 'efficient' fields, a long-term pH decrease through increasing use of nitrate fertilisers and long-term drainage deterioration through unwillingness to invest in drainage work (Curtis, Courtney and Trudgill 1976, Agricultural Advisory Council 1970). This kind of farming also leads to destruction of the resource of landscape, referred to at the beginning of this book.

This contradiction of the destruction of the resources needed in the long term is produced by the imperative of short-term profits. Really big money is made during the initial growth period in capital investment, and in general the entrepreneur does not want to invest in safeguarding long-term assets. There is a need for quick realisation of investment.

This is a form of *externalisation* (to the distant future) of costs, which is, again, an inherent feature of capitalism. The object of it for the entrepreneur is to maximise the benefits of investment by making sure that as much of the profits as possible return to the investor - that is, by internalising them. Conversely, if costs can be externalised to society, then the investor has to pay a lower proportion of them. Hence under capitalism it is good practice not to recycle, at the firm's own cost, residues from production, or to remove pollutants from industrial or agricultural wastes. If these can be 'democratically' spread, as lead in the air, or acid rain, or chemicals in the sea, then all bear the cost, rather than the producer. Neither would the technocentric solution of making 'the polluter pay' through a pollution tax work under capitalism, for the costs would then be inevitably passed on to the consumer rather than decreasing profits. If, as a result of this, demand fell, then more money would be spent in advertising the product to create more 'want' for it, or prices could be fixed through the monopolies which are an inevitable outcome of the 'free' market. Or the product could be produced in countries where there were no pollution taxes. Hence, many countries are unwilling to provide disincentives to industrial investment through unilaterally-imposed pollution taxes. In this way, through threatening to close down and reinvest elsewhere, the London Brick Company in Bedford was able in 1981 to resist moves to make it curb its own substantial contribution to local atmospheric pollution and to acid rain falling in Scandinavia (*Guardian* 21 March 1981 and Blowers 1984).

If resource depletion, growth and pollution are results of the inherent contradictions of capitalism, can it be assumed that they would not occur under socialism? Burgess's view is that this was indeed the spirit of what Marx had to say. In a socialist mode of production the conditions would be right to arrive at a harmony between man and nature (note that this does not mean that such harmony is inevitable). And this is the argument which is reiterated by Soviet academics today. Thus Ryabchikov (1976) maintains that 'An objective analysis of industrial waste shows that the atmosphere, soil and river waters of the Soviet Union are three times as clean as those in the US, Japan and the Federal Republic of Germany', because the system of state planning and economic management is more effective than private ownership in environmental protection. However, this does not entirely square with reports of grave environmental problems in Russia and the questionable ability of the Soviet leadership to combat them by enforcing existing regulations (Shabad 1979), or of the self-admitted failure of China to pay enough attention to bringing into play the 'superiority of the socialist planned economy' in coordinating economic development and environmental protection (Huanxuan 1981). But, as we suggested in section 6.2, the Marxist has a 'comeback' in his view that the economies of such countries are not in any case socialist. Thus:

> This '*a priori*' acceptance of Soviet society as socialist has a theological quality to it which leads to some rather interesting contradictions...the polluted air of the Don Basin and the black waters of the Volga...both are excellent indices of the distance between the Soviet and socialist reality (Burgess 1978).

Ensenberger's (1974) view of these contradictions spells them out more fully. He points out that the Soviet Union has, in fact, abolished private ownership of the means of production - to this extent it is socialist. However he believes that this of itself is not the ultimate conclusion of Marxist analysis. For Marxism does not see capitalism as a mere property relationship - it is a *mode of production* which leads to alienated and reified *relationships of production*. Such relationships are still found within the crudely materialistic Soviet system - they were not eliminated by the revolution which changed the property relationships, and until they *are* eliminated, that is, revolutionised, ecological contradictions will persist. (See also section 7.2.6 on naive 'ecosocialism' for other reflections on the dangers of oversimplifying Marxist perspectives.)

6.4 MARXIST CRITIQUES OF ENVIRONMENTALISTS

If we understand the basis of Marxism and how it sees nature and environmental issues, we ought, then, to be able to understand the criticisms which Marxists make about the environmental movement. We will summarise them here, and also refer readers to sections 5.3 and 5.4 on the ideological content of science, section 7.3 on ecofascism and 7.2.6 on naive socialism. All of these contain an element of the Marxist perspective in their criticisms of environmentalism, while the last also examines some shortcomings of the socialist ecocentrics themselves.

The most fundamental errors of the environmental movement generally are seen by Marxists to be its idealism (as opposed to a materialist stance), and therefore its a-historicism. Environmentalists fail to recognise the social and historical nature of resources and ecological problems. They ignore the importance of the mode of production in conditioning our perception of nature and society, and they therefore present environmental dilemmas in terms of fixed and unchanging natural limits upon human action. This determinist stance leads to fatalism and a pessimism about man's ability to create harmony with nature through enlightened social reform. This is politically reactionary, leading to repressive attitudes which are legitimated through the appeal to an 'objective' science and a pretence that environmental problems are 'above politics'. So the environmentalist ideology is deeply conservative, since it advocates measures which strengthen social 'order' and 'stability' and negates the importance of class struggle and the need for revolution. In fact, the environmental movement is a defensive one, in which middle classes and capitalist entrepreneurs are attempting to protect their interests, which are increasingly threatened by the inherent contradictions in capitalism.

The attack on a-historicism is apparent in Sandbach's (1978B) dismissal of *Limits to Growth* because this study does not place the question of resource needs and depletion in the context of economic and social organisation. It treats the ultimate fixity of resources as an urgent problem, while Sandbach does not see this as important for the near future. He prefers to concentrate on that element of current 'shortages' which are a function of commercial monopolies and of maldistributions induced by social and economic injustices. Pollution, he says, should be regarded as an outcome of private property ownership. Garrett Hardin may indeed agree with this, but his solution to the problem is to enforce compliance with anti-pollution laws, while socialism would make such repression unnecessary.

The idea of natural limits on human action is deterministic, and determinism, says Burgess (1978) 'has offered within a period of two generations intellectual nourishment for the most anti-humanitarian, anti-rationalistic tendencies in modern history'. There is a close relationship between this philosophy and fascism and imperialism. Burgess cites in support social Darwinists and their contribution to the idea of the state as an organic body, fighting for living space like an animal - the 'intellectual' justification for Nazism's expansionism. And modern biological determinists, like Desmond Morris, Eysenck and Skinner, are accused of perpetuating ideas which absorb the history of man into natural history. The 'biologising' of history, however, conceals the choices which are open to people to change their society and shape their own destiny. To make man subject to external laws of nature, independent of history, is 'to speak as though the social relativities that prevail under the capitalistic system belong to the order of nature' and this is 'to attribute to them perennial validity' (Quaini 1982). This, says, Burgess, is what the Ehrlichs do, and in thus denying the historical quality of the man-nature crisis, they construct:

> falsely universalised explanations...on the basis of the 'natural' or 'cosmic' properties of these deficiencies. Thus 'original sin', 'human nature' and the 'power of animal instincts in man' are all revived so that the crisis in man's relationship to nature can be invested with a universal significance that is ultimately determined and so the specific historical and class nature of the crisis can be conveniently glossed over (Burgess 1978).

This is convenient because it encourages people to blame themselves in a personal sense rather than the capitalist mode of production, and it encourages fatalism - a belief that 'If man is made by these conditions and the world not made by him, how can he change it: the answer is, he can't'. It is convenient to a particular group of people - the bourgeoisie - who wish to reinforce the *status quo* and protect their own class's resource interests. And, says Burgess, the 'bourgeoisie has always had the ability to generalise its own particular class view of the world as being a world view - a universal truth' (see section 5.3). In reality, this appeal to universals; to generalities which may be dubbed as 'transcending politics' in order to make us all 'pull together' is in essence highly political and reactionary, and furthering social divisions.

The ecologists appear to reject conflict in favour of a 'mass movement', and Lowe and Warboys (1978) find that this appearance has characterised other movements, like the British Campaign for Nuclear Disarmament, which have involved questions of the relationship between science and society. These authors cite many ecocentric appeals to a 'universal' ecological imperative; from McHarg's desire for social cohesion in the face of 'ecological determinism', to Mead's view that there is only one earth and that capitalism and socialism are equally bad at protecting it, to the assertions of Huxley, Fraser Darling and

Alsopp that the east-west conflict is a distraction and an evasion of the real crisis facing mankind. Thus faced with an 'external' threat we are all enjoined to bury our differences in coping with it. However, say Lowe and Warboys, this is hopelessly utopian and naive, because it fails to understand the realities of power, and to say how this unity is to be achieved (see section 7.2.5). It masks the really important question, which Ensenberger poses, as to the differences between those who travel first-class as opposed to steerage on 'spaceship earth'.

The apparent apoliticisation of the issues is carried further by those who identify them as purely technical. The issues are *diagnosed* as such by scientists and technologists, who also propose that *they* should be the main group involved in solving them. Thus the problems are fragmented and their social dimensions are denied, and, say Lowe and Warboys, this kind of science simply masks ideological sssumptions in notions of the 'facts of the matter' (see the preface to this book). It dresses environmental difficulties up as questions of resource management which rest on broad consensus, instead of seeing them in terms of the reality of a coalition of powerful property owners protecting their interests. However Ensenberger (1974), in his seminal 'Critique of political ecology' does expose this coalition, whose interests are, as he sees it, served by technocrats at all levels of the state and industrial apparatus. They are busy finding quick technical fixes to environmental problems (they are technocentrics), but they are not properly aware of the real nature of the problems and 'They can be included in the ecological movement only in so far as they belong...to its manipulators and in so far as they benefit from it'.

As to the coalition itself, it is between so-called 'concerned and responsible citizens' who are 'overwhelmingly members of the middle class and the new petty bourgeoisie', and the entrepreneur class. The former express their own 'powerful and legitimate needs' and could be a mass social force since they conceal reserves of militancy. However they are usually involved with small and limited objectives and are ideal fodder for demagogues - and the other partners in the coalition are demagogues.

In support of his thesis that middle-class environmentalists are merely defending their selfish interests in the face of threats posed by capitalism, Ensenberger points out that 150 years ago industrialisation was causing severe environmental problems, yet no predictions of doom and ecological collapse were made then. 'The ecological movement has only come into being since the districts which the bourgeoisie inhabited and their living conditions have been exposed to those environmental burdens which industrialisation brings...', he says, going on to maintain that what terrifies the doom prophets is not so much ecological decline as its universalisation. Perhaps he is thinking of Alsopp's (1972 p93) claim that 'The blame cannot be placed upon any small section of the community but belongs quite definitely to the main "consumers", the proletariat. The more we proceed towards a fair division of the spoils the greater blame accrues to the "common people"'. Thus the bourgeoisie are faced with the decline of their own quality of environment which results from increased access to this resource by the masses (which itself results from the capitalist contradiction of overproduction that makes market expansion essential in order to maintain and enhance aggregate consumption). The response will be to warn those who do not have resources that the resources will simply be destroyed if they try to get a share of them, because of the natural law of environmental capacity. This is the middle classes 'kicking the ladder down behind them' after they have mounted it (Crosland 1971) - it is the lifeboat ethic (see section 7.3).

Thus for Ensenberger 'In its closest details, both the form and content of the Ehrlichs' argument are marked by the consciousness (or rather unconsciousness) of the WASP' (White Anglo-Saxon Protestant). The preoccupation with environmental collapse is a phenomenon belonging entirely to the superstructure - that is, no real political reform of the material economic base is advocated to avert it, neither is it seen as a manifestation of that base (see section 6.1). And the concern about social and ecological collapse is really an expression of the decadence of a bourgeois society which sees its own collapse as the end of the world and wants to preserve the past rather than welcoming a future of social change. The bourgeois environmentalist's future is really a return to the past, because he wants to get rid of what he created, that is, the industrialisation to which he owes his own power. So 'the imminent catastrophe is conjured up with a mixture of trembling and pleasure and awaited with both terror and longing'.

People in this emotional state are obviously open to manipulation and they are manipulated, by the monopoly corporations: the multinationals. These powerful members of the entrepreneur class, the other partners in the coalition, are also having their interests threatened by the environmental crisis. Their resource supply is jeopardised, as is the health of their workforce. The prospect of a Third World and Communist expansion fuelled by the resources which the West has monopolised up to now can, however, be deflected by using arguments which attempt to deter developing countries from the 'folly' of growth, population or economic.

But it is not all bad news for the entrepreneur. For environmental concern can be exploited as a means of nourishing a lucrative environmental protection industry. This is 'a whole industrial complex for manufacturing and installing pollution control systems, made up essentially of those firms whose activity is at the source of the worst pollution!' (Castells 1978 p157). It includes firms such as Dow Chemicals, Westinghouse, General Electric, Rockwell, and so on. This industry helps furthermore to solve the demand problem of capitalism (caused by the contradiction of surplus value which means that the workforce does not get paid enough to be able to consume the products which it has made). It is a source of production which does not have to be satisfied by direct consumer demand, rather like 'defence' (the same firms are often involved in both industries). Instead it is largely the consumers' taxes which pay for environmental protection through the intermediary of the state.

Like the Ehrlichs, Ensenberger advocates that Western economies should contract while those of the Third World grow. This is the only way to obtain some of the benefits of Third World growth which environmentalists desire without also further degrading the earth's environment. But whereas Ensenberger, as a Marxist, recognises the deeply political implications of this equalisation (it would have to come through revolution), the Ehrlichs, he alleges, are not serious about it. They are unwilling to consider any radical interference with the USA's political system. They imagine, or pretend, that radical social-economic reforms can be achieved *through* this system, by a mixture of moral appeals to people, by letter writing campaigns and by selecting the right parliamentary candidates. In other words, like many environmentalists, they perceive the political framework of their activities as being one of a pluralist democracy. Ensenberger, with his Marxist perspective, calls this 'A crude picture of political idiocy', in keeping with the boundless ignorance on social matters which ecologists are wont to show.

6.5 MARXIST PERSPECTIVES ON ENVIRONMENTAL PROTEST

Sandbach's (1980) view of what pluralism consists of is summarised in section 1.5.2. As Bullock and Stallybrass (1977) put it, the concept of pluralism is:

> frequently used to denote any situation in which no particular political, ideological, cultural or ethnic group is dominant. Such a situation normally involves competition between *rival* elites or interest groups, and the plural society in which it arises is often contrasted with a society dominated by a single elite where such competition is not free to develop.

This contrast embodies a major distinction between Marxist and pluralist perspectives on how decision making operates in Western democracies. R. Dahl (cited in Sandbach p107) epitomises the pluralist notion that such 'democracies' are indeed democratic. He says that in them every citizen has the right to seek, and *possibility* of seeking, access to the political process in pursuit of his own preferences. Disputes (say, about environmental matters) are settled within a planning system which has a consensus of support (so, by implication, do the decisions which are reached). M. Parenti (also cited in Sandbach p107) maintains that if there are elites in this process they are *representatives* of many different interests vying for support, and the result of such competition is in line with what the majority of people wants (this view is clearly analogous with that of free-market economists on the outcome of open competition between firms). The elites, whose power is said to derive from their public support, are numerous and specialised, and they are checked in their demands by the institutionalised rules of the political culture and the competing demands of other elites. This mechanism ensures that members of the public each have equal opportunities to influence planning decisions.

We have already noted that Marxists oppose this interpretation, drawing attention to the middle-class nature of environmental protest.

The resources of money, articulacy, education and time which this group possesses may give it an overwhelming advantage over the interests of those in lower socio-economic groups when it comes to decisions about *where* some environmentally-damaging project may be carried out (see, for example, Kimber and Richardson 1974). But when it comes to questions of *whether* such projects should go ahead at all, the locus of power and advantage may, in the Marxist view, be even more narrowly defined.

For it follows from Marxist analysis that the division of labour under capitalism prevents upward mobility between classes, and maintains a non-egalitarian society where the ruling capital-owning classes are more enabled that others to realise their own interests. Structural constraints operate in favour of the ruling class and against oppressed pressure groups - the techniques of planning, for example, reflect and reinforce the social order of capitalism. Thus it is naive for environmental protest groups to appeal to supposedly neutral authorities who are set up ostensibly to balance and reconcile conflicting interests. For these authorities, such as planners or members of Parliament, cannot act as environmental managers in a way which is free from the constraint of the social-economic structure (Sandbach 1980). They are, so to speak, located in the institutional superstructure of society, and their perceptions and actions cannot be evaluated in isolation from the relations of production which arise from the mode of production at the economic base.

Although they may not class themselves as Marxist, it seems that some environmental protestors are coming round, by dint of their own frustrating experiences, to this view of the relationship between environmental planning and the economic interests of the capital-owning classes. John Thyme (1978), the inveterate anti-motorway campaigner, has spelt out how this relationship might be structured. He describes a 'technological imperative', by which environmentally-damaging technologies, such as motorways, SST aircraft, nuclear power stations or nuclear weapons, are advanced in an 'unstoppable' (and therefore undemocratic) way. There are various components of the imperative. There is the technology itself; there is a lobby on government to ensure that decisions are made in its favour at all levels, and there are scientific and technical experts to advance it - their careers depend on their support for the technology. The case is also supported by a 'brainwashing organisation, loosely staffed by hack economists who can be guaranteed, with endlessly-repeated shibboleths obscured in high-sounding technical language, to establish "economic truths" in the interests of the imperative' (Thyme p93). This apparatus all works to support the interests of a transnational industrial/financial complex, which has made its role one of researching, building, selling and using the technologies. Perhaps the existence of such a complex could still be compatible with a pluralist view of society if it could be shown that the complex was just one of a number of competing interests within a democratic framework. But Marxists would argue that this condition is not present, and that the odds are overwhelmingly controlled by and stacked in favour of the industrial/financial complex. This is also Thyme's conclusion, and he thinks that the stacking of the odds comes about largely by the establishment within government departments themselves of an interest section that will cooperate with and advance the complex's ambitions. Such a section exercises its power 'not in the national interest but in the interest of one industrial/financial lobby - in short,...it thereby constitutes a corruption of government and thus a major threat to our democracy' (Thyme pp91-2).

The importance of such interest sections in allowing the lobby to manipulate decision-making in favour of its own growth 'cannot be exaggerated. The interest section within the relevant department is crucial. Linked and cemented to the industry by bonds such as directorships on retirement...', one of its functions is to deny information and data to MPs and ministers and thereby make them more subject to the section's influence and control. 'It is in these factors that we have the essential conditions for the breakdown of our representative system of government', says Thyme. In this comment on the threat to pluralist democracy he is mirrored by Benn (1980) whose experience as Energy Minister in the 1974-9 Labour administration led him to the belief that civil service 'mandarins' exert extremely strong political influences on environmental decision making.

Thyme's idea that these influences are brought to bear in support of particular financial interests is corroborated by Hamer (1974), who demonstrated how strong is the hold which the (private) road lobby has on government apparatus, as opposed to that of other (nationalised) transport concerns. It is also startlingly demonstrated by a 'leaked' letter and memorandum on the subject of a possible public inquiry into lorry weights, which was written by and circulated among *civil servants* in the relevant government department in 1978:

> In general we welcome the idea of an inquiry as a means of getting round the political obstacles to change the lorry weights... Given that the more straightforward approach is politically unacceptable, I have no doubt that the extra time and staff effort needed...would be a worthwhile investment... For the purpose of this note it is assumed that we wish to stick to the current maximum of 10 tons axle weight...but to move, as soon as Parliamentary and public opinion will let us to a maximum gross weight of 38 or 40 tons... The main advantage of some kind of inquiry, held wholly or partly in public, would be *presentational*... At the end of the day, recommendations would be made by impartial people of repute who have carefully weighed and sifted the evidence and have come to, one hopes, a sensible conclusion in line with the Department's view... The fact finding value of most inquiries of this kind would in this case be heavily subsidiary... The above would be the position if all went well. On the other side there would be certain problems, costs and risks. The worst risk, if not the most likely, is that the inquiry would produce the wrong answer... Although the establishment in the public mind of a clear and overwhelming case on balance for heavier lorry weights is seen as the main end of the inquiry... It should provide a focus for the various road haulage interests to get together, marshal their forces and act cohesively to produce a really good case which...should do good to their now sadly tarnished public image (Transport 2000 1983).

Like Sandbach (1980 p126), we quote these documents at length because:

> there is no clearer indictment of the pluralists' emphasis upon public debate and participation as an end in itself. Indeed there is no clearer example of how public participation is recognised as a lever to support an economic interest.

This example is matched by Walter Marshall's views on the Sizewell Inquiry into building a Pressurised Water Reactor. 'I expect to get approval in about a year's time. By that time the British public will be bored to tears by nuclear power. That, of course, is the purpose of having a public inquiry' (*Guardian* 7.12.1983). Presumably it was this spirit which moved the Central Electricity Generating Board to order, in February 1984, £12m-worth of equipment for a PWR - 9 months at least *before* the verdict was expected on whether the PWR would be allowed to go ahead. These cases stand in stark contrast to the pluralist opinions of Peter Shore (Environment Secretary for the same Labour administration for whom the Civil Servants quoted above worked). In 1977 he said of public inquiries:

> first...as a result of our inquiry processes we can make better-informed decisions which fit the facts and which fit into national, regional and local objectives; second...[we can hear and] have answered the legitimate anxieties of people who have the right to express their concern; third [we can achieve] a measure of consent when final decisions are made (quoted in Sandbach 1980 p116).

While it may be fair to say that a decade ago many environmentalists would generally have agreed with Shore, and seen their job as making out environmental protectionist cases at inquiries, now the climate of opinion is far less sanguine. Even conservative ecocentric groups such as the CPRE have expressed reservations about how democratic is the 1983-4 inquiry at Sizewell. And the Oxford-based consultants, the Political Ecology Research Group, which itself gave evidence at the Windscale Inquiry in 1977, advised Friends of the Earth in 1983 to boycott Sizewell - because the group felt that the decision to go ahead with the PWR had already been taken before the inquiry began. As Ross Hesketh, a former CEGB physicist put it:

> People make a decision that they would like the facts to be such-and-such and the scientist is then sent away and asked to provide such evidence as justifies a decision already taken (LWT 1983).

Hence the experiences of environmentalists who may not have approached the issues from any overt ideological viewpoint have tended to support Marxist rather than pluralist interpretations of environmental decision making. At best the agencies and processes which are supposed to be neutral arbiters are heavily weighted towards and manipulated by the owners of capital. Through them, environmentalist protestors are put at immediate disadvantage when they try to make their cases in 'democratic' forums. They have very little money by comparison with that commanded by the industrial/commercial concern or the government department whom they oppose (and in Britain the Government is still not prepared to fund objectors at inquiries, which are carried out on an extremely expensive quasi-judicial basis). They are frequently not allowed to challenge the asserted need for a development, but only *where* it may be placed (as in the case of many motorways, or London's Third Airport). The Inspectors who preside over public inquiries will often be appointed by the same minister whose department is proposing the development - here, in effect, the ministry is judge and jury in its own case, and this situation is, in 1984, to be challenged in the European Court of Human Rights by Friends of the Earth in respect of the inquiry into the proposed construction of the M40 motorway.

At worst, 'neutral' agencies and processes are so biassed towards the capitalist class that they may engage in corrupt practices. Public inquiries, and the cost-benefit analysis which they frequently embody are pervertible, and, it is said, have been perverted to serve the interests of this class. As Sandbach puts it: 'As with commodity goods, capital owners and the ruling class are better able to ensure that the evaluation processes in environmental assessment operate in their own interest'. And the planners and government officials who operate the processes are party to an alliance between government and industry, while they deceive the environmental lobby into thinking that they are objective mediators.

The point can be illustrated by reference to environmental protection legislation in the USA. In 1969 environmental pressure groups won an apparent victory when the National Environmental Policy Act was established. This required that an Environmental Impact Statement had to be produced by any Federal Agency which proposed to carry out major development work (roads, dams, housing). The Act somewhat ambiguously told the Agencies to use 'all practicable means consistent with other essential considerations of national policy' to ensure a safe and pleasing American environment. The complicated assessment processes were to be monitored by an Environmental Protection Agency.

But as a major Federal Agency in itself, the EPA and its personnel soon became mindful of its own vested interests in growing and maintaining the political favour in which it was initially held. It soon became, in the view of environmentalists, sometimes unprepared seriously to challenge business and commercial interests. Thus it permitted developments which supported these interests even when they were supported by Impact Statements of dubious quality. And on occasions it was accused of straight duplicity - for example by issuing permits for developments before Impact Statement procedure had been properly observed. In 1983, in a scandal which was dubbed 'Sewergate' by the media, corruption charges were made against EPA officials - particularly the Agency's chief administrator, who was involved in investigations that linked the White House with business interests.

According to the critics in Congress, the EPA had withdrawn 'cleanup money' from a toxic waste site in California to prevent Democratic Governor Jerry Brown from claiming any credit for eradicating the pollution caused by the site. This was done just when Brown (who had frequently aligned himself with ecocentric groups) was running for the Senate. He lost, to a Republican who was backed up by the EPA in claims that he could and would remove dioxin pollution from a local community. But subsequently it was revealed that no method of removal had existed. EPA officials had also been accused of deciding 'not to take action against a suspected carcinogen after a private (and possibly illegal) meeting with representatives of the formaldehyde industry', of encouraging the 'Thriftway Co of New Mexico to ignore the law restricting the amount of lead in petrol', and, in one case, of involving officials who had been former employees of a waste dumping company in negotiations with that company. In all these as yet unresolved cases, the EPA, had, it was alleged, 'consistently come down in favour of big business rather than the community at large'. The White House had claimed executive privilege on some relevant documents, shredded others and erased computer records, and the ensuing row was said to encompass 'all the worst charges against the Reagan administration - cosying up to big business, dubious ethical standards, insensitivity to environmental issues and political manipulation' (*Sunday Times* 6.3.83 and *Guardian* 3.3.83).

These charges, however, do no more than reflect some of the President's own pre-election promises to sweep away much environmental protection legislation and to consider abolishing the EPA itself, thus encouraging and assisting resource exploitation by many of the commercial interests who helped to pay for his White House campaign. And the much-vaunted Environmental Impact Statement process - though it was extended to apply to state government agencies and many private firms - is generally thought to be inadequate, and open to use in the service of vested interests or as a 'self-serving agency tool' (Sandbach 1980).

In the scandal referred to above, alleged malpractices were revealed by an employee of the EPA itself. Upon making his suspicions publicly known to Congress he was, he said, harassed by his departmental head, his job was downgraded and his salary was frozen. In this way the democratic right of the employee to voice reservations about what his employer was doing was perverted. Yet this right is probably vital in any supposedly pluralistic process whereby contrary opinions about policy are theoretically able to emerge. And to assist the process there must be a great degree of freedom of information and opinion, if true democracy in decision making is to occur. The curtailment of these freedoms by actions and threats which jeopardise careers is not, however, uncommon. Over the issues of nuclear safety in Britain, for example, four scientists recently lost their jobs when they disagreed with their employers. Ross Hesketh, quoted above, lost his job in 1983 after he had written to the *Times* disapproving of a deal whereby British power stations would supply plutonium to the USA. Trevor Brown was, he said, forced to take early retirement from his Ministry of Defence job at Aldermaston when he maintained that inadequate radiological safety measurements were being taken at that atomic weapons establishment. Rodney Fordham, who opposed the safety case that was being made for the Sizewell PWR, lost his job with the UK Atomic Energy Authority (LWT 1983). Barry Matthews, a soil surveyor, was summarily dismissed from his government job in 1981 for warning people about radioactive contamination of a beach near a discharge pipe from the Windscale nuclear reprocessing plant (*Guardian* 14.11.1983). Rodney Fordham is one of several people who maintain that in the nuclear power industry it is not possible to gain professional advancement unless one is in favour of PWRs rather than other types of reactor. Certainly there are many instances from this industry, in France and the US for example, of how those of its scientists who have dissented from the industry's conventional wisdom have been subjected to career reprisals from their superiors (Pringle and Spigelman 1982). Career advancement, however, has been no problem for Walter Marshall, who in 1982 left his job as head of the UKAEA to become head of the CEGB. The latter became even more fervently committed to the use of nuclear - particularly pressurised water - reactors to generate electricity after the move - which was initiated by a government that was also committed to nuclear power generation. To a Marxist (and many others who take a materialist perspective on these matters) the inevitable links in the chain of vested interest are completed with the involvement between Marshall and the Westinghouse Corporation (manufacturers of the PWR) which the former admitted in a TV interview (Granada TV 1983).

This kind of apparent corruption of democracy may appear bad enough in itself, but an even more insidious form of perversion of the pluralist ideal comes in the processes by which protest *against* the establishment eventually becomes subverted and therefore *assimilated by* that same establishment (the latter is defined as the complex of financial, administrative and scientific personnel and institutions who together constitute and serve the interests of the capital-owning classes). The notion of assimilation was represented in its most devastating form in Orwell's *1984*, when the central character discovered that a revolutionary document which excited his subversive instincts, and was supposedly forbidden for citizens to read, was in fact written by the very authorities which it attacked.

In Bookchin's (1980) opinion, much of the ostensibly radical Marxist literature which attacks capitalism has an analogous function. He calls the Marxism of André Gorz, Herbert Marcuse and David Harvey crudely mechanistic and eclectic (see section 7.2.6), making for a 'socialist' ecology which appears to be an effective critique of capitalism (strewn as it is with red flags, clenched fists and 'sectarian verbiage'), but is in reality irrelevant *bourgeois* ideology. It is just a counter-revolutionary form of opportunistic politics, of the sort which helped people like Jerry Brown to take power in California (sitting alongside Jane Fonda, Ho Chi Minh, Fritz Schumacher and Buddha as evidence of 'les neo-anarchistes in the American ecology movement'). In other words, environmental protest in this form is really just another manifestation of the ideology of the ruling classes, because it does not, in Bookchin's view, propose and act on the revolutionary means whereby a truly different, non-hierarchical, society can be established.

Bookchin also turns his withering scorn on to the 'managerial radical' of the environmentalist movement, who displays no audacity in his protests, but instead carries his attaché case of memoranda and books into the lecture hall - he has traded off his idealism for an obsession with the mere techniques of how to manipulate opinion. In fact a radical critique of social life has now been generally lost. 'What we now call "radical" is an odious mockery of three centuries of revolutionary opposition, social agitation, intellectual enlightenment and popular insurgency'. Substituted for them have been 'the numbing quietude of the polling booth, the deadening platitudes of petition campaigns, car-bumper sloganeering, the contradictory rhetoric of manipulative politicians, the spectator sports of public rallies and finally the knee-bent humble pleas for small reforms'. The very goals of contemporary radicalism have 'all the hallmarks of bourgeois middle-aged opportunism - a degenerating retreat into the politics of the "lesser evil"'; a 'sclerotic ossification of social ideas, organisational habits and utopic visions'. In this way, thinks Bookchin, the radical ecology movement has become 'assimilated' with other protests, as a respectable movement with limited non-revolutionary ideas and demands. This completes the 'colonisation of society by bourgeois sensibility' - a result of the colonisation of society by the market, which Marx detailed. For says Bookchin with an Orwellian ring:

> the market has absorbed not only every aspect of production, consumption, community life and family ties into the buyer-seller nexus; it has permeated the opposition to capitalism with bourgeois cunning, compromise and careerism. It has done this by restating the very meaning of opposition to conform with the system's own parameters of critique and discourse (Bookchin 1980, introductory essay).

This assessment constitutes one of the most cynical repudiations of pluralism possible, and it is also very difficult to discover what positive action Bookchin himself proposes in real opposition to capitalism. But despite this, his devastating critique is indispensable for any environmentalist who wishes to achieve a heightened consciousness of the nature of his own protest and its relationship to the society in which it takes place. And heightened consciousness - of how, for example, the economic interests at the base of society become translated into a pervasive and domineering ideology at the superstructure - is an essential prerequisite of a truly democratic society in the Marxist view. Only with a developed consciousness about social relationships can we become aware of what alternatives there may be to the pervasive ideology, and the socio-economic forms which support it. This principle is illustrated by Coates (1976), in an essay on the question of how we determine what are 'needs'. To find out what true needs are, he says, there must be the articulation of need by people who are fully *aware* of what alternatives are *possible* (this is very different from the capitalist manipulation of 'demand' from a population which has become anaesthetised to social and material alternatives). However, obstacles will be placed in the way of heightened public awareness, and participation in what are regarded as either the authoritarian or technocentric preserves of need determination. These will include restrictions on the knowledge which must predicate true consciousness. Bureaucratic and political limitations will be placed on information flows, in support of the institution of private property which requires that certain kinds of knowledge are restricted to exclusive proprietorship. 'To establish', says Coates, 'truly universal access to knowledge would be to negate the dominion of resources by particular interests'. So therefore that access is withheld, and thereby the so-called pluralism of capitalist countries is dissociated.

The dissociation process is assisted by the communications media and the fields of intellectual organisation (education). It leads to the dilemma which faces Orwell's (1981 p60) society: 'Until they become conscious they will never rebel, and until after they have rebelled they cannot become conscious'.

The role of the communications media in limiting social consciousness in Britain is not in doubt, according to the analysis of TV news broadcasts which has been carried out by the Glasgow University media group (Philo, Hewitt, Beharrell and Davis 1982 p143):

> Both the press and broadcasting are active and quite consistent in their support of powerful, established interests... Information is both controlled and routinely organised to fit within a set of assumptions about how the world works and how it ought to work. The media relay the ideology appropriate to a population which is relatively quiescent, and actually promote that quiescence by limiting access to alternatives.

Elsewhere (pp128-9) the established interests are identified as essentially those of private capital. And the role of the media is to justify those interests while at the same time purveying the myth of pluralism, and the mistaken belief that what is good for business and industry is of necessity synonymous with the 'national interest' - with what is good for the majority of people. Thus the media may allow the airing of gentle modificatory dissent and promote the belief that such an airing could radically change conventional wisdom and policy. But to convey a sustained critique with a comprehensive view of what a truly radically alternative society might be like, and how achieveable, is not on (note, for example, the moves to abolish an 'alternative news' programme on Channel 4 in Britain after heavy pressure from establishment figures, and indeed to modify the political stance of the whole network after only one year of its operation).

As the Glasgow Group says (Philo *et al* p153): 'The tight central control of the BBC and the IBA, along with the power of private capital in the British broadcasting industry is unique in Europe for its non-recognition of the political and cultural pluralism of modern society'. This 'pluralism' has for long been openly flouted in and through the newspaper industry. A plurality of views is generally absent in the British press, whose owners are overwhelmingly dominated by supporters of free-market monetarism. Their nakedly-exercised power over what is supposed to be press freedom is illustrated by a comment from the managing director of *The Times* to its editor: 'The Chancellor of the Exchequer says the recession has ended. Why are you having the effrontery in *The Times* to say that it has not?' This kind of intervention was extended to the rewriting by the paper's owner, Rupert Murdoch, of leader columns which did not slavishly take a monetarist line. Murdoch eventually sacked his editor because of the political differences between them, although the former had assured the Government when he took over the newspaper that its editorial freedom would be guaranteed (Evans 1983).

If the behaviour of the media, then, does not bear out the case for the reality of the model of Western society as pluralist, what of the process of formal education, which Schumacher calls 'the greatest resource'? We shall examine this question in the final chapter.

CHAPTER 7

THE POLITICAL ROOTS OF ECOLOGICAL ENVIRONMENTALISM

7.1 A POLITICAL SPECTRUM

Not for the first time in this work, we have come up against a seeming paradox in the debate over the environment - one that could be crucial for the environmentalist movement. The paradox is that in the popular perception ecocentrics are usually seen as radical proponents of social reform, and as essentially progressive. Politically they are either seen as left of centre (as in the German Green Party) or - even better from the point of view of the image many seek (e.g. in the Ecology Party's general election broadcast in 1979) - as 'above' conventional politics and concerned with issues which transcend traditional left/right divisions. But our historical and materialist analysis of ecocentrism (Chapter 6) suggests the reverse; that it may be a (middle) class response to contradictions in capitalism, essentially conservative, reactionary, 'bourgeois' to the core and very much involving traditional political concerns. Which analysis is correct?

The answer is 'both', or 'either', depending on what particular group of ecocentrics you examine. There is a complete political spectrum, ranging from, for example, the Zero Population Growth movement to the Socialist Environment and Resources Association (SERA). As O'Riordan has said (1977), this range can be found in both ecocentrism and technocentrism, but by definition the former looks for radical social solutions while the latter seeks to accommodate environmental 'problems' within the prevailing socio-political system - either through a free-market economy approach or via gradualist liberal reforms which gently intervene in it. Therefore, on the ecocentric side the question is more stark; ecocentrism could push us towards more radical, and very left- or right-wing solutions, as Stretton (1976) has shown. The choice may be between 'ecosocialism' or 'ecofascism', by which we mean environmentalist arguments and concerns not 'above' or unrelated to traditional political concerns, but stemming from, and used very much as agents to advance, the interests of one traditional political side or the other. This issue has underlain much of what has been written so far. However, it may be useful now to tackle it in more head-on fashion, by examining the roots of the ecocentric's vision of an ecologically harmonious and viable future - of, that is, the ecological utopia, or ecotopia.

7.2 ECOSOCIALISM

The first thing we can see is that ecotopian visions provide us with nothing new, in that the geography, sociology and economics of the ideal 'ecological' society closely match those of ideal societies which have been envisaged through history - particularly by socialist thinkers. If we look at the theory and practice (particularly in the 19th century) of such societies and the communities forming them we can see at least a certain *coincidence* between ecological and socialist utopias (sections 7.2.1 to 7.2.4). This will not of itself establish the extent to which ecocentrism might in future move our society towards more, rather than less, social justice and harmony between man and nature. To ascertain this we shall also need to examine the extent to which modern ecocentric ideology - that is *intention* - has elements of socialist ideology in it (section 7.2.5).

7.2.1 Kropotkin's Blueprint for the Future

A point-by-point examination of the programme laid out in the seminal *Blueprint for Survival* furnishes us with a sociology, a geography and an economics of ecotopia, which is reinforced by Schumacher (1973) and fictionalised by Callenbach (1978). At the end of the examination we could be excused for feeling surprise at the omission from *Blueprint* of any reference to the man from whose ideas this study is so clearly derivative; the geographer, Peter Kropotkin (1842-1921). Kropotkin pre-empted *Blueprint* by 73 years, publishing *Fields, Factories and Workshops* in 1899. This work was edited in 1974 by Colin Ward, who added the word *Tomorrow* to the title to indicate the degree to which Kropotkin was a man ahead of his time. On about twenty major points there is close agreement between Kropotkin and *Blueprint* about the shape of the ideal future British society, with perhaps crucial disagreements on three. The points can be grouped under the headings: economic and social organisation; agricultural organisation and population size; industrial organisation, city and country.

Economic and Social Organisation

A central principle of both envisaged societies is the satisfaction of human needs by the minimum use of energy. To this is added the achievement of a diversity of occupations, where industry and agriculture are combined in the same communities, and where individuals combine hand and brain in their work. Man, it is said, needs to follow many pursuits rather than being confined to one, whereas the division of labour makes him servant to a machine (cf. Marx) and unable to see why the machinery performs its rhythmical movements. The ecocentric reorganisation of work would also emphasise durability of goods and craftsmanship, and jobs would be given '...on the basis that work must be provided by the community for the sake of that community's stability and not because one group wants to profit from another group's labour or capital' (*Ecologist* 1972, 17). Similarly, Kropotkin criticises measurement of the wealth of nations by the degree to which profits are produced for the few. He also looks for creativity, and advocates, like Schumacher, an economics 'as if people really mattered' - one which creates work where people live and does not concentrate them into cities, and which would not require a technology obtainable only through high inputs of capital (Kropotkin 1899, Chapter 1).

Kropotkin and the *Blueprint* both believe that international competition for trade and resources leads to war, and they advocate a degree of self-sufficiency while at the same time discouraging the myopia of nationalism. The scale of thinking has to be international yet the scale of social geographical organisation is small, with community thinking *and* global awareness. A decentralised rural-urban society, organised in small political units to be responsive to direct human contact and control is envisaged. *Blueprint* spells out the size of administrative units (p15), with '... neighbourhoods of 500 represented in communities of 5000, in regions of 50,000, represented nationally, which in turn should be represented globally'. Economies of scale and the centralised society are dismissed - the first as a mythical concept, and the second as an excuse for giant firms to dominate the market and people. Here Kropotkin departs slightly from *Blueprint*, for he believed in 1899 that large-scale society was *already* breaking down.

Agricultural Organisation and Population Size
Here is the area of most incompatibility between the two views. While they both agree on the need for self-sufficiency for Britain and on the need for cooperatives and intensive cultivation, producing market-garden landscapes, Kropotkin emphasised that because of private land ownership agricultural output was not being maximised. He thought that a change in ownership allied to technological innovation could dramatically increase food production. The *Blueprint*, by contrast, says little on ownership, and opposes high technology in agriculture. Kropotkin goes on to argue that 'If the population of this country came to be doubled, all that would be required...for 90m inhabitants would be to cultivate the soil as it is cultivated in the best farms in this country, in Lombardy, and in Flanders...' that is, ordinary 'high' farming (Kropotkin 1899 Chapter 2). By contrast, *Blueprint* (p13) says '...on the evidence available it is unlikely that there will be any significant increase in yield per acre, so that there is no other course open to us but to reduce our numbers... Since we appear capable of supporting no more than half our present population the figure we should aim at over the next 150-200 years can be no greater than 30 million...' As Ward points out, Kropotkin was wildly overoptimistic about the agricultural potential of Britain's soil (Agricultural Advisory Council 1970, Rees 1983), but he was partly right about it, as witness the nearness to self-sufficiency which was achieved in World War II. However, in his attack on Malthus, Kropotkin came near to a Marxist viewpoint on 'overpopulation' (see Chapter 6), and one far removed from that of early-1970s ecocentrics.

Industrial Organisation: City and Country

Through economic reorganisation, industry will become small-scale in these utopias, and it will be decentralised. Kropotkin thought that capitalists did not amalgamate and centralise for technical reasons (i.e. greater economic 'efficiency') - rather they did so in order to dominate markets. But small-scale autonomous industries could overcome disadvantages of market weakness by federating into cooperatives for buying raw materials and marketing goods. In the small, communally-owned village factories machines are not abandoned, but mechanisation's purpose is to free the worker for work in the fields, and to help him to concentrate on more creative aspects of his job. In other words, the technology is to be 'appropriate', and also ownable by the small community. All these chords were struck again by Schumacher (1973, 1980), and, too, in the *Blueprint*, though not so explicitly in the latter.

The implications of this theme and of urban decentralisation for the landscape are profound. As Kropotkin said, industry would be 'scattered' throughout Britain and the world and of course it would be scattered through the countryside. It is explicit in Kropotkin and the *Blueprint* that landscapes would be very different from those of today. There would be no big conurbations, separated by large tracts of highly-mechanised, large, open-field countryside, lacking people and animals. Instead, cities would be greatly reduced in size, and would have vegetation growing in the parks, streets and every conceivable place (see Callenbach 1978). There would be hamlets and villages everywhere else at frequent intervals, with small intensively-farmed plots, and the current trend of hedgerow, wood and copse removal would be reversed. The 'rural' landscape would contain more people, animals, small factories and workshops. As Ward and Sandbach (1980) note, there are marked similarities here with the idealised landscapes of Mao's China.

These ideas - derived from Kropotkin, rather than the more politically reactionary romantics - are said to have greatly influenced the Garden City movement in Britain. This represented an attempt to mix town and country, and to combine the advantages of both, and it was pioneered by Ebenezer Howard (1902).

7.2.2 Anarcho-Communism

Perhaps the really interesting question about this comparison concerns *why* the *Blueprint* does not acknowledge its debt to Kropotkin. The asnwer seems to lie in the reasons why each vision was advocated and seen as utopian. *Blueprint* (p14) tells us that it wants to create a new radical decentralised social system 'not because we are sunk in nostalgia for a mythical little England of fetes, olde worlde pubs, and perpetual conversations over garden fences, but for far much more fundamental reasons'. These are: that to deploy a population in small towns and villages is to minimise its environmental impact: that diversified (and therefore much more ecologically-stable) agriculture gives no scope for large-scale centralised farming; that the small community is the social 'organisational structure in which internal or systemic controls are most likely to operate effectively' (such controls will be necessary to help surmount 'the heavy burden on our moral courage' which the transition to ecotopia will represent); and, lastly, that 'it is probable that only in the small community can a man or woman be an individual'. Of these four reasons, only the last is not specifically centred on ecosystems principles; it alone is not essential in deferring to the 'natural law' of the mechanistic systems view of life (see Chapter 4).

Such reasons are distant from those of Kropotkin, who *mainly* wanted to create a more fulfilling society in which the individual was not frustrated and dehumanised by large-scale mass organisation, where relationships of domination and hierarchy were replaced by mutually supportive and willingly-accepted relationships between people. The ecological imperative was relatively slight.

Kropotkin's priorities were, in fact, those of an anarchist, or, as Breitbart (1979) has it, an *anarcho-communist*. And one can well imagine that the founders of *The Ecologist* (partly financed by the arch-capitalist speculator James Goldsmith) would have baulked at making the socialist roots of ecotopia explicit to a middle-class readership attracted by the thought that ecological issues were so important as to be 'above politics'. Anarcho-communism:

> ...sometimes called decentralised or libertarian socialism...rejects the contention of individualist anarchism that personal freedom is more important than, or incompatible with, social responsibility ...Anarcho-communists recognise the essential *interdependence* of personal autonomy and community. Enrichment and growth of the human personality are seen to depend on an identification of each individual with the interests of the larger group which supports personal freedom, fosters egalitarian cooperative relationships, encourages diversity and impedes the emergence of hierarchies of authority or power... The critique of existing society overlaps with Marxism in its abhorrence of repressive class relations and the inevitable inequities which flow from capitalism and other forms of economic exploitation (Breitbart 1979, 1).

Anarcho-communists, then, attack centralisation, hierarchy, privilege and domination - in capitalism and wherever they arise (Bookchin, 1980, attacks fiercely 'neo-Marxist' bureaucracy) - because they 'inhibit the development of a cooperative personality'. Anarcho-communists favour common ownership of the means of production, and distribution according to need. They hope to replace government manipulation of power with a politics which aims at the direct participation of people in daily decisions about their lives - that is, direct worker and community control.

We quite clearly have here a brand of socialism, but one which cannot be interpreted as leading to state control. Rather, it emphasises that attention to the importance of the individual which one can also find in many ecocentric writers. Anarchism goes further, however, and attributes most social ills to the domination of one individual by another - that is, to the hierarchical structuring of society. This is a road down which some ecocentrics are not prepared to go, preferring to say that if the rest of nature is hierarchical then so, too, must society be hierarchical (Goldsmith 1978; see Chapter 4 and Figure 5). However, anarcho-communists like Kropotkin, like Elisée Reclus, his contemporary, and like Murray Bookchin, a present-day anarchist, all see that the *domination and exploitation of nature by man is but an extension of the domination of man by man*. Thus, 'Both Kropotkin and Reclus...laid the conceptual foundations of a radical theory of human ecology. Ecological despoliation was seen to reflect imbalances in human relationships - domination of nature thus following from human domination' (Breitbart 1979 p2). It follows that if domineering and exploitative human relationships can be avoided in small-scale decentralised societies then such societies are also best for a harmonious man-nature relationship. The line of reasoning, then, is very different from that of *Blueprint*.

Kropotkin derived his inspiration from observing this ideal kind of social organisation 'in action' during a time spent in the Swiss Jura, where the artisans of the watchmaking industry had been substantially instrumental in creating a centre of anarchist activity, the Jura Federation. But in being thus inspired, Kropotkin was himself following the example of another: Rousseau, the 18th-century philosopher who laid down the basic social (as distinct from ecological) principle which is implicit in ecotopia - that is, that one best realises one's individuality in and through serving the common need.

Rousseau is said by Skolimowski (1981) to have been part of the 'anti-positivist' tradition in Western thought. By this is meant that he is classed as a romantic: opposed to the views associated with the development of classical science as we have described it in Chapter 2. For example, he argued against those of his fellow 'Philosophes' who supported an enlightened despotism, where there would be a strong government, planning involving the scientific expert, and a ruler with the power to put the plans into effect (such views are reminiscent of technocentrism, and were derived from Francis Bacon). Rousseau did not think that science was effecting social progress; rather, he thought that modern civilisation was regressive, and that happiness lay in a simpler existence closer to nature. He championed freedom, and thought it possible for men to be both free and to live in a political society, run by those who formed it.

Rousseau wrote about the community where he had stayed near Neufchatel. It was a simple rural community of equals in which the individual and community good came together. These 'cantons' were sovereign political societies, united in a confederation, which were small enough to have the whole adult population meet to legislate: allegedly analogous with the Greek City State (Cranston 1966).

Rousseau's essential argument in *The Social Contract* (1743) was that men do not have to choose, as Hobbes maintained, between being free and being ruled. They can be both ruled *and* free if they rule themselves. In fact, it is only through living in civilised society that men experience the fullest freedom - the freedom we have when living in a primitive state of nature is of a crude and lesser kind. The freedom of relating fully to society is nobler. This freedom results from a social contract, in which the people at large mutually agree on the desired shape of society. The laws derived from such a contract have great strength, because sovereignty derives from all the people. Rousseau said it must stay there, and must not be placed in elected 'representatives'. Although the laws can be administered by representatives, the tendency for the latter to take over the function of lawmakers must be resisted. Thus liberty consists

> less in doing one's will than in not being subject to that of another... In the common liberty no-one has a right to do what the liberty of any other forbids him to do... A free people obeys but it does not serve.

Here we have the roots of the theme of reconciliation of freedom and authority, which seem to run through to Kropotkin, to the *Blueprint* and to Bookchin (1980).

On the way, other romantics, like Shelley, Coleridge and Wordsworth, seem to have espoused the ideal of independent communities. According to Hardy (1979) they were, however, taking their cue from Godwin, Malthus' adversary, rather than Rousseau. The former, in his *Enquiry Concerning Political Justice*, envisaged a non-hierarchical, mainly agrarian, parish-based decentralised society, in which full happiness would come to individuals through defining their own rights with regard to the rights of others; but not through having imposed government.

Hardy sees a link from Godwin to Kropotkin via the French writer Pierre-Joseph Proudhon, who advocated a system of many small-scale private property owners acting as a force for decentralisation. The combination of individual ownership and mutual trading - 'mutualism' - could combat state control. Here we see a 'capitalist' element (private property ownership) mixed in with communist elements - the kind of politically inconsistent mixture which is repeated in *Blueprint for Survival*.

Hardy (pp181-7) gives two examples of anarchist communities, near Newcastle and Sheffield, operating between 1895 and 1900 to test Kropotkin's theories. The first consisted of 20 acres farmed communally by about four families living under no authority and drawing no 'wages' - any votes which were taken were not binding on the individual. Similarly, the Sheffield community had seven members living under no rules, and arranging the day's business and work in conversation over breakfast.

7.2.3. Utopian and Agrarian Socialism
The link between the kind of ideal social organisations envisaged by ecocentrics and socialists is thus seen in anarchism, particularly in 19th-century expressions. Another such link comes with parallel developments, in utopian and agrarian socialism. The utopian socialists, particularly Robert Owen, William Thompson and Goodwyn Barmby, shared Rousseau's belief that freedom was possible only in a community of equals. There was a clear goal of social change towards

harmonious cooperative communities, as a reaction against the alienation of industrial capitalism (Hardy 1979 p21). The familiar themes of decentralisation and collapse of the town-country dualism emerged. In Owen's paternalistic mind this collapse had a specific purpose - that of the improvement of people. His view that people's characters are shaped by their environment, rather than any innate ideas, harked back to the philosopher Locke.

Owen is most famed for his attempts to provide, in practice, the kind of environments which would produce rational and unselfish people. According to Hardy, after his first experiments in New Lanark in the period 1800-25, 16 American and 10 British communities were associated with him. The communities were of 800-1200 people in farming and manufacturing villages set in the centre of their land. In them, educational institutions specifically set about the 'formation of character', while the central standard of worth was human labour, on which currency was based. Both of these notions surface strongly over a century later in the mythical community of *Walden II* (but here the underlying ideology could be regarded as altogether less socialist - see section 7.3). Interestingly, Owen, like Kropotkin, thought much of Britain's agricultural potential, given labour-intensive cultivation. Owen's projection of a possible 100 million population outstripped even Kropotkin's fantasy of 90 million.

To attain utopian socialism, control of the land was needed. Hardy points out that for agrarian socialists such control was the *primary* goal. Land was held to be the source of all wealth and power, the redistribution of the latter seemed a certain outcome of redistributing the former. Hardy points out that this kind of 'socialism' was regressive rather than progressive because it simplified explanations of alienation and increased the town-country dichotomy. Furthermore, it did not necessarily lead to the true socialist ideal of *common* undivided property, even though it did follow on from centuries of steady dispossession of peasants from the land - against which there had been periodic outcries from people such as Wat Tyler, Thomas More, the Diggers and Levellers, and Tom Paine.

Thus the extent to which agrarian socialism was truly 'socialist' is questionable. For example, Feargus O'Connor's attempts to establish Chartist agrarian communities in the 1840s and '50s (he thought Britain could support 300 million people!) hinged on funding by a land company which collected money by private subscription. And Ruskin's ideas of 'feudal socialism' seem markedly hierarchical and reactionary, involving restoration of 'lost values' through returning to an economic form based on arts and crafts, and a medieval social form in which all paid a tithe to the 'Guild of St George'. This would establish a national store for the common good, but the common good involved 'no liberty upon it; but instant obedience to known law...no equality...but recognition of every betterness we can find' (from Ruskin's letters, 1871, quoted in Hardy p80). This atavistic attempt to 'recapture the wholeness of a lost order' characterises some modern ecocentrism, but it may have more in common with ecofascism than ecosocialism.

7.2.4 Cooperatives and Modern Alternative Communities

Rooted in the social philosophies of some of these writers and activists, especially Robert Owen, was a socialist movement which offered the kind of alternative economic organisation now advocated by ecocentrics such as Schumacher. That is the *cooperative* movement, which gained impetus in the 19th century in Britain and died away in the 20th - although the 1970s recession may have given it a new boost. While cooperatives may sometimes have involved a complete range of alternative lifestyles to those of capitalism, more often they concentrated on just one aspect of economic organisation, production or consumption.

Both producer and consumer cooperatives grew in Britain in the early 19th century, and there were 454 by 1863. The most successful was the Cooperative Wholesale Society, a consumer coop. Producer coops also grew until 1905 (there were 109) but then began to decline (to 18 in 1977). Beatrice and Sydney Webb were instrumental in assisting this decline. These influential socialists attacked producer coops because they allowed management to be sacked by workers. They promoted instead consumer coops. and shaped the present-day Labour Party policy which still prefers nationalisation to democratic ownership of industry by the workers in it.

However an upsurge of interest in producer cooperatives has accompanied the failure of capitalist-organised business and industry to produce employment in 1970s Britain. The publicity accorded to three attempts to create coops from failed businesses - Kirkby Manufacturing and Engineering Ltd., *Scottish Daily News* and Triumph Motorcycles, Meriden - has hidden the growth of other cooperatives. There are now some 300 associate member firms of the Industrial Common Ownership Movement, embracing a range of activities from traditional crafts to computer software. These are described by McRobie (1982) as having been started by young people 'more concerned with changing lifestyles and with the explicit social functions of enterprises than were the initial members of ICOM' (who go back into the 1950s).

This concern with the social functions and purposes of production, rather than a narrow preoccupation with its economic results, is characteristic of the cooperative movement, and it also characterises many ecocentrics, who favour an 'economics as if people really mattered' (Schumacher, Capra). Indeed, what must be the most successful producer cooperative in Europe, that based on *Mondragon* in the Basque Country of Spain, is constantly held up by ecocentrics as a shining example of how 'alternative' lifestyles can succeed.

Mondragon consists of over 60 industrial and 5 agricultural commercially viable cooperative enterprises, centred on a local cooperative bank. They were developed as an alternative to capitalist *and* nationalised production in order to defuse worker-management conflict. The workers elect a control board which hires and fires top executives. With this proviso, the latter are left to manage - they hire and fire middle management. This system accommodates the kind of concern voiced by the Webbs, and the view represented by McRobie as to why so many producer coops fail; that 'there is simply no substitute for management skills'.

Mondragon's success is thought also to be a function of the *small-scale* nature (400-500 workers) of most of the enterprises. (Campbell, Keen, Norman and Oakeshott 1977, cf. Schumacher), and of the workers' financial and social interests being linked directly to the success of the enterprises. In 1975, 58 enterprises, with a £200m turnover and no bad debts, employed 13000 workers. Associated with and supported by them were schools, a technical institute, R and D, social services, banking and housing facilities - all run as cooperatives too.

Each worker's degree of say in policy is a function of his labour and not the capital which he is required to invest, and he is not permitted to profit by selling his share in the enterprise. He receives no wages, but an 'anticipo', i.e. an anticipated share of profits - and 90% of the profits are used to finance expansion. Expansion means *more jobs*, not necessarily more machines, and the system is flexible in responding to workers' needs - if they want more leisure they can have it. While pay is higher than in the commercial sector for workers, the reverse obtains for management. Campbell *et al*, describe the latter as compensated by their youthful idealism and motivated by Basque nationalism, commitment to community and to the political idea of cooperation. The whole society is described as a set of machines and production processes run *to suit human beings, physically, socially and intellectually*, rather than having humans as slaves to machines. While social services are subsidised, they are not 'free', in order to emphasise the principle that 'each man should use what he needs, but not more, and only what he really needs'.

All this is very ecotopian, and it is also Kropotkinesque and clearly hinges on some fundamental principles of socialism. Though Britain has its cooperative movement, it has no integrated cooperative communities which emulate the Mondragon success. There is, however, a network of many very small communities, like that at Redfield (Winslow, Bucks). This consists of 20 people living off 18 acres and aiming for food and energy self-sufficiency and a legal and financial structure in which all can participate as equals. And there is the promise of the 'Third Garden City' at 'Greentown'. This is a 25-50 acre area of Milton Keynes new town set aside for the purpose of creating small-scale cooperatives based on the town-country marriage and the community control of land values and its own affairs. Greentown is founded on Ebenezer Howard's principles - themselves influenced by Kropotkin as we have noted - but it is also to incorporate the new ecological imperatives (see McRobie pp259-260).

7.2.5 The Elements of 'Ecosocialism'

There is, then, a considerable degree of coincidence between ecological visions of the future and the visions and practices of anarchists and socialists. We must now see to what degree these are *mere* coincidences, as opposed to inevitable outcomes of any essentially socialist ideology inherent in ecocentric theories. In other words, we must ask to what extent one can characterise ecocentrism as 'ecocosialist' in its roots (SERA 1980).

Bullock and Stallybrass (1977) define socialism as 'a social system based on common ownership of the means of production and distribution', and differing from communism by an 'attachment to ethical and democratic values, as well as by an emphasis on the distinction between common and state ownership'. The idea of economic growth, they say, was traditionally connected with socialism, but more recently this has had to confront 'the general realisation of the global scarcity of resources and the prospect of demographic expansion'. We have already seen, in Chapter 6, that this confrontation has produced both a challenge to the apparent absoluteness of statements about scarcity, and a reaction favouring more even distribution of such resources as are manifestly 'scarce'.

Bullock and Stallybrass also raise an interesting point about 'salvationist utopianism', which 'has always been connected with intolerance' and therefore tyranny. They say that communism has retained a utopian tradition but that socialism - a movement essentially confined to the European left - has neglected it, and this may be the root of the difference between the two. We can perhaps use this distinction also when we contrast the detailed blueprint of a future society which stems from anarcho-communism with the less prescriptive nature of ecosocialism. The latter, to be socialism, should confine itself to constructing a critique of capitalism, should spell out broad principles which should guide future socio-economic development, and should identify the importance of the mode of production in social change. (Such a confinement is consistent with the logic stemming from Marxism as a theory of liberation. If the hallmark of a future (socialist) society is that men make their *own* history, then by definition the shape of that society cannot be closely prescribed or even predicted.)

Drawing on these principles, on the Marxist critiques discussed in Chapter 6, and on the traditions of socialist communities and cooperatives, we can expect the ideal elements of ecosocialism to be as follows. Its exponents are aware that absolute notions of resource need, scarcity and 'overpopulation' can be redefined in terms relative to the culture and economy of the time and in terms of resource maldistributions between classes of people. This awareness is founded on a historical perspective in which changing modes of production are recognised to be important in influencing changing attitudes to nature. So the perspective is not based on explanations derived from 'natural law', 'human nature' or other crude and authoritarian determinism. Ecosocialism's left-wing radicalism criticises existing forms of social order and enunciates principles underlying new forms - principles which eliminate alienation, state control and centralisation, and narrow nationalism. Alternatives to present forms of production would be planned on the principle that private profit is unimportant compared with social and environmental justice and wellbeing.

We will now unpack these socialist elements a little, and see whether they are at all present in ecocentric writing. (If they are, then this fact may at least partly repudiate the Marxist criticisms of ecocentrism described in section 6.4.)

Redefining Resources and 'Needs'
The idea that what constitutes a resource is not absolute but is relative to the definition of need, and that the latter cannot be divorced from need creation in a capitalist society has been discussed in section 6.3. Ecosocialism would be concerned to recognise this and to redefine 'needs' in such a way as to de-emphasise the philosophy of materialism so essential for capitalism's continuation. There is indeed a tradition of such recognition in ecocentrism, ranging from Thoreau ('Most of the luxuries, and many of the so-called comforts of life, are not only not indispensable, but positive hindrances to the elevation of mankind', *Walden*, p22 Collier edition), through to Schnaiberg. The latter, following Galbraith, sees artificial need-creation as a vital part of capitalism's production treadmill, which must be reversed, in his view, to stop environmental degradation (Schnaiberg 1980, pp228-9). The recognition is partly explicit in *Blueprint's* belief that it is possible to redefine needs in a stable-state society in order to emphasise, in a radical alteration of social relationships, the enhancement of creativity and spiritual fulfilment. Thus, the 'intensity of relations with a few (people) rather than urban man's innumerable superficial relationships - this is the compensation for decreasing stress on consumption'. The theme carries through into the current political manifesto, which has pages on the importance of the arts, values and 'the spirit' (Ecology Party 1983).

Redistributing Resources
At the same time, ecocentrism seems to show some awareness of the need to redistribute resources - particularly between Western and Third Worlds. *Blueprint's* (p13) estimate of an optimum global population of 3.5bn rests on assumptions which include 'that there is absolutely equitable distribution, no country enjoying a greater *per capita per diem* protein intake than any other... Utopian though they may be, unless these assumptions are realised, we are faced either with the task of reducing world population still further until it is well below the optimum, or with condoning inequalities grosser and more unjust than those which we in the developed countries foster at present'. Similarly, Ehrlich and Ehrlich (1972) incline towards the socialist economist Pierre Jalee's view that underdeveloped countries are having their natural wealth plundered by developed countries and that 'de-development' by the West and Third World development are prerequisites to global ecological soundness. They say (p443) 'Redistribution of wealth both within and among nations is absolutely essential'. Again, this view comes through in the conclusion of the *Limits to Growth*, in Schumacher's *Small is Beautiful* and in the modern Green Parties. The Ecology Party declares that the much-discussed Brandt Report identifies relevant Third World problems but comes to the wrong conclusions, failing to discuss the need for land reform and to curb the present role of Western multinationals, e.g. in repatriating

profits. The Party makes notably little of 'overpopulation', an obsession of its predecessors (see section 7.3). Indeed, it says 'Shortages of food are not the primary causes of world hunger - nor is population growth', and it calls for voluntary population control but alleges that compulsion and coercion are unacceptable infringements of human rights.

Egalitarian relationships of production are also wanted for the West - there are calls for equal pay, equal rights for women and racial minorities, equal job opportunities and a national income scheme with a minimum wage. These calls, however, stop short of demands for common ownership of all of the means of production. Thus, while land is regarded as common wealth, and a common heritage not to be exploited, and 'The monopoly control of wealthy landowners and institutions must be brought to an end', private ownership of 'houses, farms, workshops or anything that improves the usefulness of the land would continue', even though 'it should not be possible to own the land itself - people should be the tenants of the land rather than the owners' (Ecology Party 1983 pp13-14). Finally, explanations of inequality and an uncaring society which reside in 'lack of funds', 'weaknesses in the welfare system', or just 'human nature' are resisted.

The Mode of Production
Much of what ecocentrics call for amounts to modification of the prevailing mode of production, but the extent to which this is recognised and is a radical call is questionable. Nonetheless, the inevitable decline of the current mode of production is foreseen. 'The principal defect of the industrial way of life with its ethos of expansion is that it is not sustainable...unless it continues to be sustained for a while by an entrenched minority at the cost of imposing great suffering on the rest of mankind' (*Blueprint* p2). And, while he is an idealist rather than a materialist, Schumacher also recognises that 'one of the most fateful errors of our age is the belief that the problem of production has been solved. This illusion...is mainly due to our inability to recognise that the modern industrial system...consumes the very basis on which it has been erected' (1973 p16).

Alienation and Good Work
The small-scale, community-based, self-reliant society is a fundamental feature of the ecocentric's utopia, and we have discussed the coincidence between it and socialist and anarchist social forms. But there is also the question of a coincidence of *motives*. At least some can be discerned.

One is that this social form is most conducive to a nonalienating form of production. In it, people are not alienated from each other, or from their task of appropriating nature for socially useful production, or from nature itself. It becomes possible to organise work in a more individually satisfying and rewarding way and to make it more enjoyable and its products worthwhile (Schumacher 1973, 1980, Ecology Party 1983, and the *Blueprint*, which believes that people should not be forced to choose either 'jobs or beauty').

Clearly there is a common desire between socialists and ecocentrics for a new social form based on a form of production that has the purpose of removing the 'cash nexus'. It will be an 'alternative production', which not only eliminates the production line but also emphasises socially useful rather than useless and damaging, if profitable, products. Ecocentrics make much play of such plans as were presented by the Lucas Aerospace Combine Shop Stewards' Committee to its management in 1976. These demonstrated in detail how production could be diverted from weapons to products which would help to solve social problems in Western and Third Worlds, and would make minimal demands on resources and improve environmental quality (Elliot 1977). Despite management's rejection of the plans, the theme of 'alternative production' has become a central plank of trades union policy in Britain (TGWU 1983), reacting to escalations of the arms race, and it links directly with the Ecology Party's call for small businesses and cooperatives devoted to community projects, and the breakdown of large defence concerns.

Decentralisation
All this presupposes forms of industrial and social democracy, which are entirely compatible with the traditional socialist opposition to a centralised state. 'It would therefore be sensible to promote the social conditions in which public opinion and full public participation in decision making become as far as possible the means whereby communities are ordered' says the *Blueprint* (p14), perhaps fumbling towards a liberal rather than truly radical solution to authoritarian state control. And, quoting Aldous Huxley, Schumacher (1973 pp27-8) argues that ordinary people should be given the means of doing 'profitable and intrinsically significant work, of helping men and women to achieve independence from bosses, so that they may become their own employers, or members of a self governing, cooperative group working for subsistence'. This would lead to 'progressive decentralisation of population, of accessibility of land' and even of 'ownership of the means of production'. Ten years later the demands are still being made, by the Ecology Party, for decentralised government (with devolved regional assemblies as a first step), a decentralised health service (with small community hospitals) and community-oriented education. These will have to be achieved through a planned economy, in which research and development, transport, energy provision and the like are coordinated through overall strategies. Quite whence the strategies will come is unclear. In an anarchist utopia they would be decided by individuals in a federation - in ecotopia they might well be imposed from the top of a hierarchy. Socialism and planning (and nationalisation) may be coincident, but that does not mean that planning cannot also be found in non-socialist totalitarian regimes - it is not an inherent socialist feature.

Internationalism
This brings us back to what might seem a contradiction of ecocentrism, which might also be held to occur in socialism. Founded on the rights of

the individual and his identity with a small local community, reorganised society nonetheless has to operate in an international context. The Ecology Party invokes us to 'act locally - think globally. That should be the watchword of politics today...we need new ideals of international responsibility and cooperative endeavour'. This, like the socialist international, may be more easily desired than achieved. Indeed, rather than having international cooperation between members of the same class, regional alliances between different classes, based on the needs of capital and reinforced by narrow nationalism seem to have perverted socialist thinking in the West (Harvey 1982 p420).

7.2.6 Is it Naive Socialism?
Despite these aspects of 'ecosocialism', the radical ecology movement still suffers in the 1980s from some of the political naiveté which Ensenberger recognised a decade ago (see section 6.4). Then, for instance, we had the neo-Malthusian appeal to be 'above politics' which in reality was so politically reactionary:

> The old grumble about the rich exploiting the poor is insignificant by comparison with the present reality of mankind as a whole exploiting mankind along with the rest of nature (Alsopp 1972 p93).

Now, despite its apparent leftward movement, we have the Ecology Party (1983 p20) with: 'in today's world it's not a question of some ending up as winners and some as losers. We're all in it together', and 'The old-fashioned politics of class conflict are grinding to a halt. The politics of life start here' (1983 p5). Here is still the negation of the importance of class struggle which, with a corresponding negation of the pre-eminence of the mode of production in influencing social consciousness about the man-nature relationship, so alienates Marxist socialists, appearing to them politically repressive and idealistic. In other words, these 'ecosocialists' are still not socialist enough.

But even those who do begin to talk in terms of capitalism being 'to blame', and of a corollary that removing capitalism would eradicate ecological problems, may attract the charge of naiveté. For Ensenberger considers these simplistic stereotyped statements as too abstract. Marxism, he says, does not exist just to produce eternal verities. Such verities merely encourage us to forget our concern over concrete ecological problems, and to be passive about them because we believe that nothing can be done short of total social revolution:

> Since the concrete problem in hand...[such as the dying rivers] can, without precise analysis of the exact causes be referred back to the total situation, the impression is given that any specific intervention here and now is pointless...reference to the need for revolution becomes an empty formula, the ideological husk of passivity (Ensenberger 1974 p19).

These verities are pure ideology for ideology's sake, such as are mouthed in the Soviet Union, where they have also become discredited by the chronic pollution problems which in reality have arisen. Russia's problem, thinks Ensenberger, is in the crudity of its materialism, which has dispossessed private capitalists while 'all other relationships remain alienated and reified'. Consequently there has not been a revolution in 'social production', that is, in the social relationships between men and nature and in the mode of production.

The enviromentalist left - particularly in the form of André Gorz and Barry Commoner - is also attacked by Murray Bookchin, who champions a 'social ecology' which appears rooted in anarcho-communism. Bookchin claims that 'environmentalism is merely a technocratic attempt to contain ecological disruption within the framework of capitalism' (i.e. it is merely technocentric). *Social ecology*, on the other hand, contains very radical philosophical and cultural implications, centring on the 'non-hierarchical nature of ecosystems (sic) and the importance of ecological diversity for stability'. These implications point to a non-hierarchical and diverse society as the prerequisite to an ecologically harmonious man-nature relationship (there appear to be parallels here with Capra's prognosis).

Bookchin sees Commoner and Gorz as vulgar Marxists, preaching a technologically-oriented form of socialism, and therefore 'environmentalists' rather than 'ecologists'. While he believes that Marxism itself is crudely materialist and mechanistic, he thinks that in any case Gorz is a 'bad Marxist'. For Gorz (1980) asserts that capitalism is capable of assimilating 'ecological necessities as technical constraints', and adapting the 'conditions of exploitation to them'. In other words, Gorz thinks that capitalism can deal with environmental problems, though he also attacks capitalism for its environmental impact. Commoner, meanwhile, is savaged as a 'closet Euro-communist', committed to centralised economic planning. Both are accused of seeing ecological problems as economic - of practising economic reductionism; a 'naive form of Marxist socialism'.

Gorz also stands accused of 'pilfering' Kropotkin's ideas in an eclectic way, without reference to the latter's broader intellectual pedigree. And he is charged with being muddle-headed - he writes of his dream of a society without bureaucracy, yet one whose 'major industries' are centrally planned. He tacitly accepts Marxism's base-superstructure model, which Bookchin rejects as reductionist (crudely deterministic) and as having been discredited by the critical analysis of Max Weber, who showed that the base and the superstructure are 'too interchangeable to be distinguishable'. Gorz's inconsistencies are, too, revealed in the way he sometimes embraces both Marx *and* Malthus and Hardin - the last with his neo-Malthusian restatement of 'original sin (i.e. the "population problem" begins in everyone's bedroom, not in the world's brokerage houses)'.

Much of Bookchin's critique is a mixture of personal abuse, along with justifiably withering ridicule of Gorz's utopian scenarios which smack of the naiveté of the *Blueprint* or Callenbach's Ecotopia - as Bookchin says: 'a childish "libertarian" Disneyland'. But inasmuch as Gorz and Commoner typify the ecocentric left, Bookchin's exposé makes us question the whole basis of this strand of environmentalism.

He poses the devastating questions which prick the bubble of the European Green movement with its pluralistic approach of reform through Parliament:

> Will ideas become a matter for serious concern or mere radical chit chat? Will revolution be the lived experience that literally provides the substance of life or entertaining and expendable episodes?

Ensenberger's challenge to the left, however, is that it may, in unmasking the ineptitude of the ecologists' social analysis, imagine that there is *no* truth in their ecological prognostications. This could be a fatal mistake, born of a failure to recognise that there has been a fundamental quantum leap in the environmental threats which are posed by modern industrialisation. Ecologists are naive and utopian, but unlike the left they do realise that 'any possible future belongs to the realm of necessity and not that of freedom, and that every political theory and practice, including that of socialists, is confronted not with the problem of abundance but survival' (Ensenberger 1974).

This kind of 'realisation' is apparently what motivated the German activist, Rudolph Bahro, to make his much-trumpeted move from socialist into Green politics in 1979, relinquishing his previous adherence to strict Marxist tradition. As a latter-day guru of radical ecocentrics, his position may be said to typify Green ecosocialism, effectively severing it from the kind of ecosocialism proposed by SERA. Bahro says that while the 'impulse to self-destruction' (via an arms race and an ecology crisis) is partly rooted in European industrial capitalism's development since 1750, the 'impulse' goes back beyond capitalism into 'human nature itself'. Thus 'so long as we continue to see class struggle as the key to the contemporary crisis we will only remain trapped in the very circle out of which it is imperative to break'. Bahro wants to shift the focus from the attack on capitalist forms of industrialisation to industrialisation itself. The impulse to obliteration lies in the very foundation of industrialisation, and we actually have too much work and too many workers. 'The abolition of at least half the work now performed in industrialised countries must take unquestioned precedence over the demand for full employment within the industrial system'. And it all apparently starts with the individual - 'We have to embark on a psychological revolution that starts with ourselves, and liberates our politics from the aggressive model of reactive class antagonism that only reinforces and accelerates exterminism'. We must agree on means and aims 'in a common project capable of subordinating the opposing special interests of all those engaged in it to their own fundamental and long-term interests' (Bahro 1982A).

Bahro claims that this is not anti-Marxist (though it seems very much to be so), for 'Marx and Engels specifically recognised the possibility of a historical situation leading to the common ruin of the contending classes'. He concludes that 'The struggle to overcome the ecology crisis takes precedence over a class struggle...' which has no perspective of superseding industrialism. Marxists today, he maintains, must 'circumscribe ecologically the traditional political economy of both capitalism and socialism...' and bring together 'the most diverse alternative attempts in thinking and living, a movement that attains a degree of cohesion and agreement such as was reached in the past only through the claims of religion' (Bahro 1982B).

Some socialists might regard this last not merely as naive but also as sinister in its potentially reactionary nature. For a quasi-religious appeal has connotations of an appeal to irrational faith, and to elements which are above and beyond human control. It is in fact more the hallmark of an ideology which is antithetical to socialism - that is fascism. Socialist and anarchist commentators have recognised fascist ideology in ecocentrism, and we shall now go on to identify where it might be. Where it is found, it often coexists with elements of its opposite; ecosocialism. This coexistence, as we noted at the beginning of the chapter, is partly because of the diversity in the origins of ecocentrism, and of its still rather muddled political state. But it is also due in part to the highly indeterminate future which lies before all of us - a future which could go, politically, either way in response to the ecological and social stresses which have become apparent. For it does not seem axiomatic, as Meacher (1976) thinks it is, that future serious resource shortages would mean that Western industrial production 'would have to be systematised for prior social uses' and that socialist and ecological perspectives would necessarily 'dovetail'. In fact, the two could drift poles apart. This much was hinted by Marx, who saw no historical *inevitability* about socialism - just that it was a *desirable* outcome of capitalism. However, there was and is a choice of 'socialism or barbarism'.

Ensenberger fears that the latter may be the outcome of the continued externalisation of the consequences of commodity production from the firm to society in general, by which environmental deterioration becomes widespread. Far from a mass consciousness developing, from which a saner society would emerge, liberal illusions could be shattered. 'Fascism has already demonstrated its capabilities as a saviour in extreme crisis situations and the administrator of poverty.' In the atmosphere of panic induced by ecological disaster, the ruling class will have recourse to fascism, for 'The ability of the masses to see the connection between the mode of production and the crisis...and to react offensively cannot be assumed'. Instead, internal imperialism will probably increase, mass consciousness will be increasingly colonialised, and external imperialism will 'regress to historically earlier forms - but with enormously increased destructive potential' (Ensenberger 1974). In this apocalyptic vision of competitive (nuclear) wars over raw materials, Ensenberger, ironically, finds common ground with one of those whom he spends so much time in demolishing - the ecologist Paul Ehrlich (1969).

7.3 'ECOFASCISM'
7.3.1 The Elements of Ecofascism
In section 7.2 we traced the roots of anarchist and ecocentric utopias back to Rousseau, and his ideas of individual fulfilment through social

fulfilment. However, in seeing Rousseau as inspiring anarcho-communist or socialist thinkers, we take a sanguine view of someone whose 'politics' have been the subject of much controversy. A less sanguine view suggests that he has inspired those at the very opposite end of the political spectrum. Writing in 1944, for example, Russell (1946 p660) called Rousseau:

> the father of the romantic movement, the initiator of systems of thought which infer non-human facts from human emotions, and the inventor of the political philosophy of pseudo-democratic dictatorships as opposed to traditional absolute monarchies. Ever since his time those who considered themselves reformers have been divided into two groups, those who followed him and those who followed Locke...the incompatibility has become increasingly evident. At the present time, Hitler is an outcome of Rousseau, Roosevelt and Churchill, of Locke.

With hindsight we might think Russell's denunciation of romanticism as particularly coloured by the undoubted romantic input into German Nazism. Nonetheless, he reminds us that the *Blueprint's* model society with its Rousseauesque flavour, could be more politically ambiguous than we have interpreted it up to now. We thus need to ask if ecotopia is as 'free' a society as ecocentrics might make out, or whether there lies within it the seeds of Russell's right-wing 'pseudo democratic dictatorship'. This is the implicit thrust of many left-wing critiques of earlier ecocentrism, such as those by Neuhaus (1971) and Ensenberger (1974), but Bookchin is more direct in talking of 'ecofascism'. He says:

> The most disquieting direction followed by many environmentalists has been in the direction of what I would bluntly call ecofascism...the message of the 'lifeboat ethic'. To utilise ecological dislocations as a means for reverting to an 'ethics' of crass egotism, to build a 'strategy' of self sustenance on the myth of a stingy nature that faces depletion of its bounty, to elicit a meanness of spirit on the presumption of 'scarcity' is an horrendous mockery of ethics, nature and even the traditional concept of scarcity as a stimulus to progress (Bookchin 1979 p22).

In this view, 'ecofascism' is the adoption of the lifeboat ethic (see Chapter 1), whose basic tenet is that because of absolute constraints on the amount of resources which are available there cannot be equality of access to those resources unless the numbers wanting access are greatly reduced. And the responsibility for eliminating 'overpopulation' lies with those who are 'causing' it by breeding too fast. Any attempts by those possessing resources to redistribute them to those who are still 'irresponsible' enough to 'overbreed' are themselves irresponsible, for they merely result in more overbreeding.

This is Malthus restated by the neo-Malthusians, most notably Garrett Hardin (see Chapter 4), and Bookchin is one of those who wish to expose its ideological content (see Chapter 5). He says that it is the ideology of *self*-interest, and *self*-preservation, thus, like a Marxist, he seeks explanation for aspects of ecocentrism not only in 'the conflict between humanity and nature...but...the conflict within humanity itself'. The lifeboat ideology will lead to an 'ethics of repression and totalitarian control'.

This repression, according to Bullock and Stallybrass (1977 p228) will consist of several elements, if we are to take the word 'fascism' literally. It has come to imply extreme nationalism, anti-communism, anti-Marxism and anti-liberalism, anti-democracy and anti-parliamentary parties. The fascist state is authoritarian, one-party and with one charismatic and dictatorial leader. All in it share a cult of violence and action, exalt war and, with their ranks, salutes, uniforms and rallies, give their parties a para-military character. They are also racist, and draw support from those on the extreme right. They attract youth, through the cult of action, and groups in the middle and lower middle classes who feel threatened by inflation, economic depression and the organised working class. Originally radical in their demands, they often shed them when they reach power. Bearing in mind Nazi Germany and Russell's criticisms, we might also add that we would expect to see in ecofascism elements of emotionalism and romanticism wedded to the above, to produce a belief in a quasi-mystical unity of people and their land, of the state as an organism, and of the 'naturalness' and therefore rightness of hierarchy, of struggle and survival of the fittest (i.e. the organic analogy and Social Darwinism).

We now intend to examine 'ecotopia' to see if it does contain elements of any of the above, and then to consider the lifeboat ethic as a recurring theme in ecocentrism. We shall also refer in more detail to the theme of racism in connection with zero population movements, and the use of science to legitimate this ideology. The argument will be that elements of 'fascism' or extreme right-wing ideology are indeed latent within ecocentrism.

7.3.2 The Unacceptable Face of Ecotopia

We have said that the seeds of socialism may be found in 'Ecotopia' (Callenbach 1978). The mythical country has decentralised production, where government control over the population is small, where the people at large decide on what technology will be developed and where workers own the means of production, and the direct investment of capital in an enterprise by absentees is not permitted.

But the seeds of fascism, some have argued, are also there. The way that power in the mythical state is seized by the 'Survivalist' party in the first place is essentially anti-democratic. (Ecotopia is in Oregon and Northern California, which is supposed to have broken away from the rest of the US and to have established an ecological state in the 1980s and 1990s.) The regular political structure is 'paralysed and supplanted', and this process is helped by an engineered political and economic crisis producing unemployment and hardship. Consequently measures can be taken which would be impossible in a pluralist democracy. Intervention by the rest of the US is prevented by the simple and 'plain nasty' (O'Flinn 1978) expedient of planting nuclear mines under New York and Washington and threatening to blow them. In addition a special militia is trained 'to resist American invasion' across Ecotopia's closed borders, and to bring about the sort of 'end to immigration' which the *Blueprint* advocates. This militia is one arm of the repressive apparatus of a superficially liberal state in which, however, the opposition party offers no real opposition. There is underground dissent, but its influence is circumscribed by Ecotopian 'counter-intelligence' which has a habit of calling on potential dissenters at night and issuing threats and intimidation. This all reminds one of the *Blueprint's* contention that the transition to its ideal society will 'impose a heavy burden on our moral courage' and will 'require great restraint. Legislation and the operation of police forces and the courts will be necessary to reinforce this restraint...' (*The Ecologist* 1972 p14). The Ecology Party also calls for a 'self-policing community', which idea is reflected in the Ecotopian practice of having a 'bad practice' list, violation of which carries the risk of 'heavy moral persuasion' and eventually ostracism by the community.

To follow an extremist ecological programme, the Government (led by the charismatic President Allwyn) jeopardises economic and social welfare, and encourages sustained debate about such vital issues as whether bulldozers should have steel or wooden fenders. Meanwhile there is a host of new laws *à la Blueprint*, which expropriate property, enforce the sorting of garbage, imprison polluters and use force to break up big corporations. The inculcation of a 'socially more responsible attitude to child rearing' (*Blueprint* p14) and of the idea of 'man's proper place' in the environment (Ecology Party 1979 p10) is via a hierarchical and élitist school system based on 'tribal groups', with a curriculum heavily laced with the biological imperative. Indeed there is, throughout, a romantic quasi-mystical sentimentality about the land, reminiscent of Hitler's Germany and based on the wilderness and 'mother earth' (with tree worship and totem poles) as well as aggressive nationalism. President Allwyn gives 'clarity strength and wisdom' and is as much a religious leader as a politician (Callenbach 1978 p38). This spills over into the cult of violence represented by the ritual war games. These semi-religious rites involve the venting of aggressive impulses by painting oneself with blood and spearing members of the opposite 'team' to maim or kill them. The populace delights in 'plain animal strength' when thus displayed by humans, and virility and ennoblement are attained by the spilling of blood. This is pure fascism, and it stems from Social Darwinism, in which (Ecotopian) man is seen as basically a tribal animal; naturally aggressive and competitive. There is, too, an acceptance that different races 'cannot live together in harmony' via the institutionalised racial segregation of 'independent' black enclaves within Ecotopian cities.

All these traits contain elements of classic fascism, but O'Flinn has drawn attention to the fact that the very existence of Ecotopia is in itself an example of the lifeboat ethic - that 'meanness of spirit' which Bookchin places at the centre of *eco*fascism. For one of Ecotopia's first problems is how to deal with the big food surpluses produced by this richly-endowed corner of the US (which also boasts more doctors and skilled workers *per capita* than the rest of the country). The answer is to run down production and stop exporting food and lumber. This is done while the rest of the US experiences chronic shortages of these commodities. Ecotopia thus offers 'not even socialism in one country, but a kind of freedom in a bit of one country, a kind of freedom that within its own boundaries is ecologically deeply sensitive but is quite ready to secure itself by turning the rest of the continent into a nuclear desert' (O'Flinn 1978).

Ecotopia, however, is but one manifestation of the self-interested 'ethics of crass egotism' which represent ecofascism. There are many other examples of the lifeboat ethic in ecocentrism. These may be drawn from 'environmentalist' movements and campaigns in the West, such as the anti-airport protests in Britain (Pepper 1980). But the most notable comments on ecofascism have usually surrounded the question of population control and Third World development.

7.3.3 Population and the Lifeboat Ethic

A cornerstone of ecocentrism in the 1960s and '70s was the appeal for global population stabilisation and even decline in the interests of controlling economic growth, pollution and resource depletion. Many saw 'overpopulation' as the most fundamental cause of environmental degradation. The naiveté of this view was, as we saw in Chapter 6, exposed by Marxist critics. The political self-interest of the appeal to do something about this 'unprecedented common danger facing mankind' (IFR 1972) has also been remarked on by a wider range of critics (e.g. Chase and Simon).

First, the more avid zero-population-growth school can be seen as advocating fundamentally repressive measures. 'We must have population control at home, hopefully through a system of incentives and penalties, but by compulsion if voluntary methods fail' (Ehrlich 1970, Prologue). Ehrlich's programme would include changing tax laws to penalise child-bearing, making birth-control instruction mandatory in schools, and breaking off relations with the Vatican. According to Chase (1980), others (e.g. Burch and Pendell 1947) advocate the adoption of laws to sterilise compulsorily 'all persons who are inadequate, either biologically or socially' and to encourage voluntary sterilisation for those who have 'had their share of children'. Simon (1981 p324) tells us that involuntary sterilisation of blacks and mental defectives in N. Carolina is already carried out, ostensibly 'in the best interests of the public at large', while the Director of the US Population Office publicly advocated, in 1977, sterilising 25% of women in the Third World, to protect US commercial interests.

Secondly, those 'environmentalists' who pin so much faith on population control are seen either to have dubious motives themselves, or to be the unwitting abetters of those with dubious motives. Simon believes that a big shift took place in official US Government policy on population control (from anti- or neutral to pro-) in the late 1960s, and that this shift was facilitated by apparently environmentalist groups, though it was not really undertaken for the sake of the environment. Instead there was, and is, thought to be a complex of self-interest motives in the thrust for ZPG (Zero Population Growth). These include the motive of numerical fear - that the continued supply of the provisions and resources of the West's 'lifeboat' of survival, many of which come from the Third World, will be threatened by Third World population growth and resultant home demand for them. As Buchanan (1973) put it, population control programmes inspired from the US 'represent an attempt by the centre (the metropoles of the White North) to ensure the continuing availability of the resources of the periphery (the ex-colonial dependencies)'. The motives also, believes Simon, include the fear of communism, because Third World population pressure often leads to civil unrest, which can produce revolution, and this hardly ever favours US interests, but leads frequently to communism. Thus, 'In every area where the Red penetration is most successful...population pressure is severe and increasing (it is)...one of the most potent factors in the success of the Reds in their campaign for world domination' (Stuart 1958 p9, quoted in Simon 1981).

These motives are allied to, and partly a function of, nationalism. Economic and political self interest are exerted through the lifeboat principle, applied to the poor/rich division between the world's peoples. The line of reasoning is neo-Malthusian, and is best enunciated by Garrett Hardin (see Chapter 1). The 'Commons' (the pool of global resources, including food) can be destroyed by 'overbreeding' and 'overdemand' on its carrying capacity. Such 'overbreeding' is basically reckless and irresponsible, and while some countries may be persuaded or coerced into adopting population control policies, others will not. To the latter we must deny food aid. We in the lifeboat must not - cannot - extend the arm of charity to these particular drowning people. For if we do we will merely encourage more 'overbreeding'. We will remove the natural check of carrying capacity, and everyone will be worse off. This is, of course, exactly Malthus' reasoning for denying institutional aid to the poor, and it leads to the view that:

> we should facilitate, instead of foolishly and vainly endeavouring to impede, the operations of nature in producing this mortality... Instead of recommending cleanliness to the poor, we should encourage contrary habits. In our towns we should make the streets narrower, crowd more people into the houses, and court the return of the plague... But above all we should reprobate specific remedies for ravaging diseases; and those benevolent, but much mistaken men, who have thought they were doing a service to mankind by projecting schemes for the total extirpation of particular disorders (Malthus, *Essay* 1803, 2nd ed., Book IV, Chapter 5).

Carrying on with the medical imagery, Hardin produces a 'triage' scheme to guide Third-World aid policy. The West should concentrate its aid to countries making progress in population control. It should abandon those in two other categories; first, those who will in any case solve their problems for themselves, and, second, those countries which are 'hopelessly' overpopulated. To send food to the likes of India, Egypt, Haiti, is to throw sand in the ocean. Ehrlich, similarly, proposes terminating food aid to countries where 'dispassionate analysis' shows that the food/population imbalance is hopeless. His views on the problem in India are summarised in a famous passage in *The Population Bomb*, where he describes his fright, on returning home one evening to his hotel in Delhi, at encounterting teeming noisy crowds in the streets. He experiences the noises and smells and observes dirt and starvation and declares that these give him an insight into the 'feel of overpopulation'. (As Chase remarks, similar scenes in London two centuries previously, had given Hogarth the feel of *poverty*.) Ehrlich, a biologist making social judgements, thus concludes that 'too many people' are at the root of social and environmental problems, and for him people become a 'pollutant'. The denigration of people in this way (or at least *some* people, for presumably Ehrlich does not think that he *himself* besmirches the earth) is seen as another element in ecofascism. Chase makes a direct parallel with the Nazis:

> There is nothing in biology in general, or in genetics and population genetics in particular, that gives a young, healthy, well-paid American in Palo Alto, California, the moral, scientific or political right to...urge the termination of food, medicine and all other aid to India and various other countries whose birth policies offend him...genocide remains genocide, whether advocated in a Munich beer hall in 1920 or in a Texas college auditorium in 1967 (Chase 1980 p402).

But Ehrlich was but one of a number of well-fed and healthy well-paid Americans who took this line. Together with other notables like Isaac Asimov, Justin Blackwelder, Paul Getty, William Paddock, and of course Garrett Hardin, he formed an organisation known as 'The Environmental Fund'. In 1976 and 1977 this published and advertised statements to the effect that food aid violates the principle of carrying capacity by 'artificially allowing more people to live on the land than can live from it'. It advocated stopping any foreign aid programme which encouraged population growth, stopping illegal immigration into America and balancing legal immigration with emigration. 'Tightening *our* belts', said the Fund, 'and "ascetising" *our* diets to solve the world food problem is the *wrong* solution', for this merely encourages more growth:

> Our past generosity has encouraged a do-nothing policy in the governments of some developing nations...we must not permit our aid to underwrite the failure of some nations to take care of their own...undernourished women are *less* fertile than well-nourished ones. Thus, improving the nutrition of poor women *increases* their fertility...simply sending food assistance to hungry nations, or even helping them grow more food, isn't enough. It simply makes the problem worse. (The Environmental Fund 1977).

The Malthusian reasoning is again evident, and in the conclusion which is reached by this organisation, (and similar organisations such as ZPG, the Campaign to Check the Population Explosion, Planned Parenthood/World Population) the numerical fear surfaces:

> With finite resources, an exponentially increasing population means an

exponentially decreasing amount of resources per person. It is not logically possible to reconcile this mathematical fact with the current wave of world wide 'rising expectations' and clamour for a 'new economic order' which will bring affluence to all. Put in a different way, this can be extended to the statement that there are no 'have not' nations - only overpopulated nations... As the world population grows exponentially, will we rapidly become a 'have not' world? (*Idem*)

Thus, questions of social injustice and resource maldistribution are denied. Instead we are encouraged to feel hostile to the (overbreeding) poor, and to worry about whether we will join their ranks. Such fears were institutionalised rapidly. In 1970 President Nixon signed into law a $400m expansion of the family planning service, and the creation of a Federal Population Office. Now the US Agency for International Development (AID) spends $100m/annum on world population control, according to Simon, in order to prevent rapid population growth in the Third World because it 'encourages political and civil disorders' and 'accelerates the use of natural resources'.

7.3.4 Scientism, Scientific Racism and Behavioural Engineering

Critics of the ZPG ecocentrics often remark on what they see as the inflammatory terminology which they use. Phrases like 'population bomb', 'population explosion', 'flood of people engulfing the earth', 'survival of the human race at stake', 'the population plague', are rife. Simon suggests that they have overtones of violence ('bomb', 'explosion') and contempt for humans ('people pollute'). This very language is an expression of 'ecofascism'. For those who believe that people pollute do not usually think this of *all* people, especially themselves. As Harvey (1974B) noted about the dangers of overpopulation rhetoric, it encourages one to think that someone, somewhere, is redundant, but to ask 'Is it me?' and answer 'Of course not'; 'Is it you?', 'Of course not'. Then the inevitable conclusion is that 'It must be *them*'.

The rhetoric, therefore, lays the ground for mistrust and contempt of other people - of their rationality, responsibility and ultimately of themselves. The concern with overpopulation has extended from mere *protective* self-interest to an *antagonism* to other groups. And, like Malthus' ideology, this ideology is sustained by a veneer of scientific objectivity and appeal to natural and inevitable law. So this is racism (and class discrimination) bolstered by *scientism*. The doctrine's (see section 5.3.2) modern function is to create a climate in which those in the lifeboat can view with equanimity their own unwillingness to help anyone else aboard.

Hardin's views on how to prevent a population disaster do not stop at the idea of birth control for all. Instead he advocates both negative and positive *eugenics* ('the production of fine (esp. human) offspring by improvement of inherited qualities' - Concise OED). The argument is that, as population growth proceeds, the proportion of 'responsible' to 'irresponsible' people will diminish, because the former restrain their family sizes while the latter do not. To correct this, there must be negative eugenics, via sterilisation of the extremely unintelligent. 'Observation has shown that, almost without exception, two feeble-minded parents can produce only feeble-minded children. There seems to be little danger of society's being deprived of something valuable by the sterilization of all feeble-minded individuals' (Hardin 1949, quoted by Chase). At the same time positive eugenics, encouraging the intelligent to have more children, is recommended. The parallel between these ideas and those of the Nazis about the need to foster growth of the inherently superior Aryan race (partly through the negative eugenic of slaughtering Jews) hardly needs to be stressed. At the basis of such ecofascism, of course, is the contentious belief that intelligence, 'responsibility' and other desirable qualities are primarily inherited rather than a function of environmental factors. Hardin had such a belief, and thinks it is sustained by analogy with experimental data derived from animals. He cites the work of Burch and Pendell, of whom the former was, according to Chase, prominent in the American Coalition of Patriotic Societies, whose objective was to prevent Nordic Americans from being replaced by aliens or negroes through immigration or breeding. Burch is said to have advocated 'scientific birth control' among 'ignorant, diseased and poverty-stricken families' and the compulsory sterilisation of Americans drawing Federal relief.

By contrast Ehrlich appears to be a moderate. He says (1970 p120) that intelligence has both genetic and environmental components - there is an inherited possible range of intelligence, but where we lie within that range is environmentally determined. While agreeing with the Hardin theory that over several generations those at the lower end of the genetic intelligence scale could outbreed those at the upper end, he finds no evidence that 'such drastic differential inbreeding exists'. However, Ehrlich's inspiration, William Voght, does not seem to be any more moderate than Hardin's antecedents. Voght, in a much-quoted book 'Road to Survival' (1948) linked population and pollution, advocated non-help to countries in the Third World, and came out with such gems as 'One of the greatest national assets of Chile...is its high death rate'. Like Hardin and Malthus, Voght attacked doctors for 'misguided' good intentions (keeping people alive against the natural law). Like Ehrlich he believed that 'overpopulation' would lead to war, and he did not think that the US should subsidise unchecked spawning in India and China by buying their goods. In his futuristic story 'Eco-Catastrophe', Ehrlich (1969) makes heavy use of the rhetoric deplored by Simon. He talks of overpopulation leading to the 'end of the ocean', 200,000 corpses in Los Angeles, and New York smogs, the formation of a 'midwestern desert' in the US, and eventually thermonuclear war triggered off by the struggle for resources.

Simon deplores the role of Ehrlich and other scientists in investing the notion of 'overpopulation' with a veneer of objective truth. In reality, he says, 'Whether population is now too large or too small, or is growing too fast or too slowly, cannot be decided on scientific grounds alone. Such judgements depend on our values...' (Simon 1981 p332).

The real motive for this scientism includes, he thinks, racism, and it is not simply directed against Third World populations. He alleges (p322) that the number of state supported birth control clinics in the US is related to concentrations of black people in various states - 70% are in 10 southern states (containing 19% of the total US population) because of a 'desire to reduce the fertility of blacks'. Whether this statistic is really to be interpreted in such a sinister light is, of course, debatable, as are many of Simon's other assertions.

The idea of selectively manipulating human development for social ends does, however, often surface in literature beloved of ecocentrics. Another much-vaunted ideal future community is *Walden Two*, conceived by the behavioural psychologist B.F. Skinner. To many ecocentrics this has come to be regarded as a kind of utopia, but to those on the left of the spectrum it is seen more as a dystopia. The particular piece of scientism which is its driving force is behavioural engineering. 'When a particular emotion is no longer a useful part of a behavioural repertoire', says the prime mover of Walden Two, 'we proceed to eliminate it... It's simply a matter of behavioural engineering... The techniques have been available for centuries. We use them in education and in the psychological management of the community' (Skinner 1948 p93). This would seem to be an excellent method of 'inculcating' the 'correct' environmental values - which is the mission of education in the *Blueprint's* world (see Chapter 8). In it, ethical training is completed by the age of six, via a system of behaviour-dependent sticks and carrots which are used in a manner compatible with the Pavlovian dog experiments. It seems closer to manipulation of the developing mind through brainwashing, and it involves a very mechanistic view of human beings. This, then, is another example of the denigratory attitude towards humans which has characterised a considerable amount of ecocentric thought, and which has come to be regarded as a major element in 'ecofascism'.

7.4 CONCLUSION

If we are to draw any conclusion from all this it must be that we have shown that ecocentrism is politically most ambiguous and that it has, as O'Riordan (1977) pointed out, distinctive and opposite political wings. We have also, it seems, supported O'Riordan's contention that *within* specific 'ecocentric' groups and individuals there are deep ambiguities and contradictions. Nonetheless, some broad patterns of change over time also seem to emerge. We noted in Chapter 1 that there has been an apparent leftward drift by environmentalists like Commoner and Allaby over the short space of 10-15 years. And even over the four years from 1979-83 the British Ecology Party has shifted to a position increasingly compatible with ecosocialism. Very important, too, is the recent appearance of books on the environment which highlight the centrality of the mode of production (Schnaiberg 1980) and the analysis of social change (Sandbach 1980).

Perhaps this is the most significant aspect of this drift. It could consolidate and reinforce theoretical backing for what is apparently happening through *Die Grunen* in Germany, the *Partito Radicale* in Italy, *Ecolo* and *Angalev* in Belgium and the *Politieke Party Radikalen* in Holland. All these groups are giving political expression within the existing system to the seeds of a radically alternative and opposite mode of production and philosophy. Whether this dialectical process will continue or whether such opposition will be assimilated into 'the establishment' and even used to justify a repressive politics remains to be seen.

CHAPTER 8

CONCLUSION: IS EDUCATION THE 'GREATEST RESOURCE'?

8.1 THE ENVIRONMENTALIST'S PANACEA

The idea that education is 'the greatest resource' (Schumacher 1973 p64) pervades both technocentric and ecocentric thought. The former argues - along the lines of the Baconian creed - that for us to avoid ecological malpractice and to 'manage' our way out of difficulties which our production system creates we must know more thoroughly the laws of nature. On the other hand the ecocentric sees that education provides answers to more fundamental problems: 'We are suffering from a metaphysical disease and the cure must be metaphysical. Education which fails to clarify our central convictions is mere training or indulgence' (Schumacher, *ibid*). So education can provide the panacea for our ills; whether we think (see Preface, section 1) that if only people knew the 'facts' of environmental abuse such abuse would stop, or whether we believe, with the *Blueprint*, that a complete change of *values* must accompany total reorganisation of society along more ecologically sound lines. Schumacher argued that education is the 'greatest resource' because it maintains and strengthens not only human daring, initiative and constructive activity but also regard for nature. If this book is to conclude constructively, by indicating the best way forward to a more ecologically and socially harmonious society, it must now examine this ecocentric claim that the best way is provided through education.

In fact the broad environmentalist consensus on the importance of education conceals a considerable division about the crucial question of what form environmental education should take. Huckle (1983) distinguishes education *about* and *from* the environment from education *for* the environment. Only the last, he maintains, offers the theory and practice with which to improve environmental wellbeing. Both the former work actually to *sustain* current deterioration in environmental quality - they help to maintain an unsatisfactory environmental, social and political *status quo*.

Education about the environment is a technocentric response to our environmental predicament. It involves the 'accommodators' - the environmental managers and planners - who use techniques from welfare economics and systems analysis in a neo-classical scientific and economic approach. O'Riordan (1981B) believes that this is the common form of environmental education, but though it has bred a new sub-species of academic environmental scientists its solutions involve no new politics and economics. Huckle says that it views environmental education and management as neutral instruments of social policy, it is formal and largely concerned with knowledge of 'facts' about the environment. Any consideration of values relates only to the 'pragmatism of the wider society'.

Education from the environment is compatible with that ecocentric thought which argues for a new morality based on ecological pragmatism combined with bioethical regard for nature. In other words, it is a moral and ethical (values) education of the kind which Schumacher and Skolimowski (1981) propose - education in not only *how* to perform technical feats, but in *what* ought and ought not to be done. This education, says Huckle, argues that environmental imperatives should impel us to forget political differences. 'In the tradition of Rousseau and others it employs environmental studies as a rationale for pupil-centred, topic-based, learning which often reflects a rather naive respect for children and nature.' It tends to ignore socio-political factors, emphasising consensus in the face of a 'common' universal threat of impending crisis. It holds not only that field study provides cognitive skills, but that such contact with nature also aids personal growth and moral development. Huckle, in line with left-wing critiques (see section 6.4) rejects education from the environment as reactionary, because of its very 'apolitical' stance.

He advocates education for the environment, because it increases pupils' awareness of the moral and political decisions shaping their environment, and helps them to form their own judgements and to participate in environmental politics. It is issue-based and it involves projects culminating in community action. Huckle calls it a 'radical' education, but we should add that it is radically to the left (since it is clearly based upon a Marxist perception of the importance of historical and materialist perspectives). Schumacher's environmental morality is also radical education, but radically to the right, since it involves the *implantation* of specific values - no matter that these values might be regarded as 'better' than those which underlie the conventional wisdom. The distinction is vital, for it is one between education as *indoctrination* and education for the development of the pupil or student as a relatively *autonomous thinker and learner*, who is moving through his life towards the independent exercise of his own reason. As Hales (1982 p16) puts it: 'the latin, *educatio*, means "to bring out", whereas administrators [and, we might add, ecocentric 'missionaries'] understand education as putting *in*, instilling certain skills, traits, inhibitions and information'. Harris (1979) calls it the difference between *imposition* (of ideology) and *consciousness-raising* (about alternative ideologies).

This book's perspective is that the kind of environmental education which would function as a great 'resource' would be that of the 'bringing out' type. It would help pupils to understand how people 'make themselves and their environment in a two-way interaction with nature which is strongly conditioned by the kind of lives they lead' (Huckle p108). It would show how human labour can increase or decrease earth's carrying capacity and environmental quality, and how attitudes to nature relate closely to prevailing economic forms (cf. Marx). Above all, it would foster critical awareness of how conventional values sustain and are sustained by a specific set of relationships of production, and it would encourage an open-minded scrutiny of alternative forms of society and sets of values. It would thus question fatalistic (deterministic) theories of history and human behaviour. So education would encourage pupils to believe in their capacity for *self*-determination. Huckle says that this should be achieved practically, by allowing them to determine their own school environment - by making the school 'a just and ecologically sound community'. However, this last point may make us a little sceptical. For the image of independent school 'republics'; self-determining and impervious to external social pressures and influences, just does not gel. Huckle rightly advocates the importance of a practical as well as a cerebral form of radical environmental education, but both he and O'Riordan (1981B) are fully aware of the current extreme difficulties of attaining either. One can sum up the difficulties by saying that society at large actively discourages that kind of educational approach. Thus Apple (1982 p130) has to admit that while the lived culture of schoolchildren constitutes a potent area for consciousness raising, any isolated attempts to explore this area through setting up democratic socialist alternative curricula and models of education 'will remain inconsequential probably'.

No educationalist can afford to neglect the fact that there is, in reality, and must be, a close relationship between what happens in education and what happens in the rest of society. Teachers operate at the 'superstructural' level, and they must take into account the relationship between what they do and the material 'base' of society. When we remember this relationship, we must argue that education *of itself* is not and cannot be the 'greatest resource'. That is to say, *it alone* cannot constitute the instrument of a fundamentally enhanced social consciousness and a change of values. Although 'Many well-meaning people have looked to education to bring about a better world...it is the height of naiveté to think that education can take on a promotional role and drastically change the social order itself' (Harris 1979 p183). It will accomplish nothing very radical unless its reform is accompanied by socio-economic reform. Indeed, there is no shortage of evidence to suggest (as Huckle concedes) that some current 'education', far from being an agent (a resource) for reform, is the reverse. It increasingly sustains and enhances the political *status quo* and those who benefit from it.

8.2 EDUCATION IN OPPOSITION TO SOCIO-ECONOMIC CHANGE

The first way in which this happens is through an omission. Education frequently *fails* to encourage critical awareness and an ability to think in new and creative ways. It does this by emphasising, often mindlessly, the techniques of *how to* do things. But it neglects consideration of values and morality. Hence it does not encourage pupils and students to question received and conventional wisdom. Whereas 'true wisdom', according to Schumacher, is *essentially* concerned with developing values about what is morally right or wrong; most education - particularly science education - merely gives us a 'tool box' by which we can perform technical feats. The second law of thermodynamics is just such a tool box, says Schumacher, denying C.P. Snow's assertion of the importance of knowing this law in order to be 'literate'. For one misses 'nothing' if one does not know the second law, but a knowledge of Shakespeare, for example, is knowledge of how to live.

The science curriculum (and most education *is*, either explicitly or implicitly, *science* education because it propagates the methods and philosophy of classical science) does not consider 'how to live'. In fact it shuns most questions of how and why the 'crunch decisions of a complex and technological society' are made, as Ryder (1983) discovered when investigating teenage attitudes to nuclear power and the Three Mile Island accident. He found that though children actually *want* to discuss the pros and cons of nuclear power, they are *prevented* from so doing by a curriculum that focusses only on 'agreed laws and theories of science', at the expense of studying controversial issues. The (technocentric) assumption is made that a 'sound knowledge of basic physics' is needed before nuclear issues could be debated, and when the curriculum is thus 'dominated by a serialist building-block approach...the length of the series is too long to reach the important decisions either before children leave school or before they are alienated from science'. In fact, science education, up to and including higher education, focusses on 'fact' gathering and rote-learning, making students puzzle solvers within a paradigm rather than investigators of the paradigm itself. And if opportunities for critical reflection should come, science students may be disoriented by them and want to reject them on the grounds that they do not constitute safe, 'hard' knowledge (Pepper 1983). This blinkered outlook may be reinforced by a curriculum that places little importance on modes of expression - written or spoken - so that the student who does want to explore values and opinions is frustrated because he cannot convey them. In this regard, classical science education seems to have taken on the functions and attributes of modes of education which it replaced, for we find that Bacon attacked medieval schools and colleges on the grounds that 'Everything is found adverse to the progress of science; for the lectures and exercises there are so ordered that to think or speculate on anything being out of the common way can hardly occur to any man'.

Such speculation, about the *purposes* of science, for example, may be discouraged by the science teacher, or relegated to a position of secondary importance, on the grounds that the scientist's first duty is to be 'objective'. Thus we find that an environmental educator at Plymouth Polytechnic berates environmentalists for having allowed moral, philosophical and emotional considerations to 'cloud' their thinking. They should complete the processes of observation, recording and experimentation before interpreting what they see. Such a scientifically correct procedure can be achieved only as long as they 'keep a detached view of their personal environmental opinions' (Roxborough 1983). Naive as this may seem in the light of what Chapter 5 of this book has discussed, it is common enough in academic circles to hear scientists relegating moral issues to the status of common-room opinion - to be aired and considered in isolation from their professional work. Harvey (1974A) categorises this false (and impossible) separation as a device whereby the corporate state reconciles its need for flexible and adaptive minds with its inability to abide free, creative individualism. It makes values a matter of opinion for the spare time, while 'facts' constitute the legitimate object of the academic's professional pursuit.

Harvey goes on to consider the idea that education serves the *status quo* not merely by its errors of omission. In a more active way, the dominant ideas in education are, he says, the ideas of the dominant (capitalist) group in society. This group is represented by the corporate state, which is a 'tight-knit hierarchically-organised structure of interlocking political, legal, administrative, financial and military institutions'. The structure is used to *transmit* information down to individuals about what is right for the 'national interest' (really the interests of the ruling class). And education is one of the means of transmission. Like the corporate state which patronises it, education is dominated by the ethics of rationality and efficiency, and it produces a technically efficient bureaucracy whose aim is to enhance the stability of the corporate state - through enhancing competition and economic growth, through managing cyclical crises in the economy and through containing or defusing discontent.

This last may be achieved as we have suggested, negatively by failure to stimulate in students any disturbing critical reflective thought about social organisation (i.e. political and social awareness). But it may be reinforced more positively by teaching - often in the guise of 'value free' science - certain values which support capitalist ideology. Frequently those values enter, unchallenged, the subconscious mind as part of the 'hidden curriculum'. Schumacher (1973 pp71-2) brought into the open those implicitly-stated ideas which he believed most children learn in Western society without realising it - or without realising that they are *only* ideas and not tablets of stone. We described them in the Preface, section 2; they are the values of Social Darwinism, materialism, moral relativism and positivism. As Harris (1979 pp140-2) describes them, these values - of hierarchy, competition and role conformity; of acceptance that the rewards of life will be unequally distributed; and that such rewards are external material things rather than intrinsic satisfaction - constitute a series of beliefs instilled in virtually all educands of a capitalist liberal democracy. This is because education in a *class* society (which it is) is a political act, 'having as its basis the protection of the interests of the ruling class'. It is a manipulation of consciousness which instils a particular way of seeing the world, in order to secure prevailing ideologies and social relations and make them free from threat. Harris' Marxist perspective thus reflects that of Harvey. But it is supported, too, by non-Marxist ecocentrics like Schumacher, and also Skolimowski, who relates the values of science education to the needs of the materialist, 'consumptive' society.

Hales believes that in its structure and its hidden and overt messages, the schooling of children in Britain functions as a template for capitalism. School work, he says, is organised in patterns identical to those of industrial work, with its hierarchy of expertise. Science figures massively as a product - received consensual knowledge - and not as a process which mediates an active reading and writing of the world. What passes for learning is 'alienated reproduction', where children work in 'transactional' exercises to regurgitate supposed free-standing truths, and where personal knowledge - of feelings, opinions or experiences - is discriminated against in favour of a supposed need to give the 'right' answers. As in the world of industrial work, school labour is fragmented, standardised and routinised (and this may be said of undergraduate education, and, indeed, of most postgraduate 'research'). This organisation, thinks Hales, is a form of social control; classes are ruled by the needs of task-attention - automatic tasks like comprehension tests in English and computational exercises in maths and science. Isolated question answering is the predominant mode of classroom work - the more so now, with microcomputers as teaching 'aids' - and teachers are turned into 'classroom managers', overseeing mundane transactional writing about factual questions. Thus, imagination, discovery and values-exchanging and -development are all stilted.

This kind of education reproduces social identities, with the teacher as a 'production line supervisor', ensuring that externally-imposed (curriculum) goals are carried out in a standard and fragmented way, and that 'knowledge', reduced to the status of a commodity, is consumed. And the 'fundamental relations of capitalist production, as routinisation, mechanisation, fragmentation, commodity production, are built in to knowledge and identified with the experience of learning and "self development" in school'. This leads to the hegemony of one social group and the subordination of others. It is a dominance which does not involve repressive force, for the dominant group stays so by virtue of the subordinate marginal group's failure to find alternative ways of working or thinking (Hales 1982 pp142-3).

Hales' theme echoes Bowles and Gintis' (1976) research, which showed that at the bottom of the American educational hierarchy the personality characteristics which were rewarded - acceptance of subordinateness, docility, punctuality, conformity - corresponded to those required for the lowest-paid jobs. Higher up, *different* characteristics commanded the rewards - such as self-control, flexibility, initiative and ability. These authors conclude (p56) that 'the major aspects of the structure of schooling can be understood in terms of the systematic needs for producing reserve armies of skilled labor, legitimating the technocratic-meritocratic perspective, reinforcing the fragmentation of groups of workers into stratified status groups, and accustoming youth to the social relationships of dominance and subordinacy in the economic system'. Apple (1979), too, highlights the *form* of the curriculum as geared to capitalist ideology. With its increased emphasis on individualised learning, for example, based on self-paced mastering of learning 'packages' (perhaps computerised) the opportunities for cooperative interaction among children are minimised.

The ideology of curriculum *content* can be illustrated by the extremely successful way in which one view of economics (free-market monetarism) has been presented in recent years, both by 'education' and the media. As Benn has noted (*Guardian* 8 August 1983), the deterministic message that capitalist 'laws of economics are inexorable' has been so firmly implanted that no alternative ways of interpreting the world have received really serious consideration. 'Even liberal progressives get hoodwinked into thinking they are facing "harsh realities"' when they assimilate the monetarist assumptions, says Benn. So it is hardly surprising when manual workers (like those at the Hyster fork-lift truck plant at Irvine, Ayrshire, see *The Guardian*, February 17 1983) actually vote themselves wage *cuts* at the management's suggestion, believing that this course of action is in the long-term 'national interest' and that there is no alternative to it.

Any education which fosters such attitudes is likely to be supported by representatives of business, commerce and manufacturing. And when such people make the familiar complaint that education is not geared to 'national needs', they really mean that they want it to foster attitudes supportive of *their* world view, and deterministic beliefs that there is no alternative to the satisfaction of *their* needs as a priority. (They may also argue that there is an objective need for 'technically trained' personnel, but this consideration is really very subsidiary, as is shown by research commissioned by the Department of Education and Science. This demonstrated that employers seek to hire the graduates of particular universities, and are quite oblivious and unconcerned about the technical expertise which their employees have acquired (*The Times*, 4 November 1983).)

Since, in 1979, a government was elected which strongly articulated the interests of the owners of the means of production, centralised and direct influence on the ideas in the curriculum in Britain has been quite openly sought. And we can see quite clearly that this influence is to be used to enhance the socio-political *status quo*. Thus, in 1982, the National Economic Development Council (supported by business and trades unionists) recommended that, to serve the 'needs of an industrial manufacturing nation', schools examination boards should be more open to influences from outside education, and that the allocation of funds - along with control of the curriculum itself - should be more centralised and more determined by the views from industry (*Guardian* 10 November 1982). Central control over funds has now been strengthened by rigorous restraint on all local authority spending and, in the case of non-university higher education, by the establishment of a central controlling and funding body whose main members are drawn from industry and commerce.

School curricula for those up to the age of 16 are also to be centralised, and government ministers will have 'unprecedented control over what and how children are taught'. History teachers are concerned that this 'nationalisation' will make their subject a tool for building nationalism and inculcating right-wing views - bringing back the discredited chronological and heroic approaches which emphasised kings and queens and battles, and Britain's part in world affairs as a 'glorious success' (Walker 1983). In the science curricula, the education minister has already rejected proposals for a sixteen-plus examination until he approves of their broad outline. He will not allow this outline to have, as was intended, 15% of physics teaching on the 'social and economic issues which arise from the application of scientific knowledge' (i.e. precisely what Ryder, quoted above, says the curriculum needs). And in chemistry and biology there must not be examination questions on 'pollution control, problems of inadequate world food supply, harmful effects of pesticides, and excessive use of fossil fuels, nuclear fuels and alternative energy sources'. And the aims of history must be 'to understand the development of shared values which are a distinctive feature of British society' (*Sunday Times*, 18 September 1983).

The kinds of values which are to be a compulsory basis for the (hidden or overt) curriculum were revealed in plans informally drawn up by ministers in 1983. They wanted to promote 'self-respect and a sense of individual responsibility' (to whom?) among the British people, such as would, for example, encourage mothers not to go out and seek a paid job. They intended to support schools with a 'clear moral base' and to encourage (i) a change in social values so that 'wealth creation' becomes more acceptable; (ii) new ways of developing children, 'the country's major resource of the future, into self-reliant, responsible, capable, enterprising and fulfilled adults'; (iii) identification of 'characteristics of behaviour and attitude which the Government might legitimately hope to see adults possess or, conversely, avoid'. And they hoped to 'make the curriculum more geared to industrial needs' (*Guardian*, 17 February 1983). Three months after developing this 'bizarre vision of social engineering' (Fairhall 1983) these ministerial 'godfathers' were re-elected to power with a huge parliamentary majority.

Although the examples cited above are of rather naked manipulation of the curriculum for economic class interests, it should not be inferred that the correspondence between education and society at large is necessarily or customarily so crude and deterministic. Social and cultural influences must not, of course, be denied. And even when economic influences on education are uppermost, the power of capital may be highly mediated. Furthermore there is not necessarily any *conspiracy* to make education serve class interest. As Apple (1982 p160) says, when describing how the American curriculum may be dominated by prepackaged material that defines the goals, processes and outcomes of learning with precision and without reference to the teacher, such material is published with state adoption policies in mind - and those policies are influenced by industry's desire to make its case the fundamental problem schools are to face. But 'conspiracies to eliminate provocative or honest material are not necessary here. The internal working of an educational apparatus...is sufficient to homogenise the core of the curriculum'.

8.3 WHAT TO DO?

The foregoing discussion emphasises the closeness of the relationship between education's content and method and the relationships of production at large. But we should recognise that this relationship is dialectical rather than deterministic. It therefore allows the possibility of a measure of 'independent' educational reform and innovation. Indeed Bowles and Gintis warn socialist educators against sitting back to await the 'inevitable' demise of capitalist economics, or simply trying to create 'islands of socialism in a sea of capitalism'. Teachers should 'attack' broadly and continuously by pressing for democratisation of educational institutions and processes; they must reject authoritarianism and be in the forefront of a move to create a 'unified class consciousness'. Harris, too, following Friere (1972), emphasises the need to raise the consciousness of the 'masses', and to attack false consciousness, though this approach should avoid indoctrinating children with the teacher's own particular world view. He holds teaching *method* to be crucial - the right methods will encourage autonomous learning and remove power relations in education. The teacher will be *an* authority but not *in* authority. And 'partial' solutions, which discover the faults of the existing system without proposing remedies, or which show people that they do not after all know the answer to questions which they had thought they could answer, are not to be eschewed. Harris makes specific suggestions on education for consciousness raising: through recognising ideology for what it is when it is articulated; through stripping away mystification; through exposing the internal contradictions of received wisdom; and through exposing false consciousness. This last is also clearly identified. For instance, explanations of the fact that there are rich and poor people in capitalist societies constitute *false* consciousness if they proclaim that God willed the situation to be so, or that there is a natural hierarchy, accurately reflecting ability and effort. But the explanation that such relationships are necessarily built in to capitalism constitutes *true* consciousness because, given our present state of knowledge, it is the more 'progressive' - it is more in our best interests rather than serving to legitimate the interests of the ruling class; it is falsifiable though not yet falsified, indeed it at present accounts for observed phenomema without promulgating illusions about the world.

But despite his call for educational reform, Harris (p182) concludes, as we do, that education as provided by capitalist liberal democracies 'will not change from providing structured systematic distortions of reality' until it is at least accompanied by raised mass consciousness promoted through 'anti-educational' means. New *lived* ideologies must be formulated - not through formal education but in general social *praxis*. Harris is speaking of new *social* and *economic* goals for communities, involving new relationships among people, as the real motor for wide social change. He is therefore reminding (ecocentric) educational reformers that people will not change their values just through being 'taught' different ones. They will behave outside the classroom as society demands and encourages them to behave. O'Riordan's (1981B) example of why gardeners and farmers will continue to use chemical sprays against their better judgement explains why we all may continue to behave, in present society, contrary to what our developed environmental awareness might decree:

> the individuals involved are subject to a pattern of circumstances which are so pervasive and complex as to be unshakable unless (a) society radically alters its ethos toward natural phenomena and the use of modern science and technology, and (b) the dominant economic and regulatory institutions are restructured to be much more sensitive to long-term ecological damage...'

What, then, is the real way forward, if it is not to be solely or even largely through education? It must be through seeking *reform at the material base of society, concurrent with educational change*, otherwise any effects of the latter will be ephemeral. Such reforms must, to be ecologically *and* socially acceptable, be along socialist lines (for reasons explored in section 7.3). And we *must* think in terms of ecological as well as social acceptability, for we must heed the warnings of environmental scientists. Our rightful scepticism about the 'objectivity' of scientists must not blind us to the high quality of much of their work and the reality of the environmental threats which it discerns, even though such threats must be seen as historically produced. Of course the biggest threat concerns the ultimate environmental disaster of nuclear war (Peterson 1983, Turco *et al* 1983, Ehrlich *et al* 1983), so there is no purely academic quality to the urgency with which new social solutions are needed.

But we should not naively imagine that a sudden heightening of collective social and ecological consciousness - stemming perhaps from resource shortages or a nuclear accident - would provide a revolutionary proletarian breakthrough to ecotopia. The more likely outcome of such a jolt would be to feed a repressive ethic of the lifeboat, or a yet more secretive and policed society where control of resources and technology was restricted to an ever-tinier privileged elite (see Harrison's (1982B) horrific vision of an overpopulated New York in the year 2000).

Instead, it is more reasonable to envisage the evolution of an improved socio-economic system from the seeds of that alternative economy which has *already* been growing steadily out of the West's economic depression in the past decade. These 'seeds' comprise: first, a network of alternative small communities where people try consciously to live along *Blueprint*-style principles; second, a regrowth of producer cooperatives, sometimes socialist, sometimes not; and third, a growing quasi-barter system where goods and services which the conventional economy cannot cheaply provide are exchanged between ordinary (often unemployed) people.

Collectively, these elements are a reaction to capitalism's failure to satisfy the material and spiritual needs of many people. More and more are becoming involved in them as unemployment grows and social welfare cushioning becomes less adequate. Often these people have capital, in the form of severance payments, which could be used to fund alternative modes of production for socially useful goods, like that at Mondragon (see section 7.2.4). Mondragon's economics - aimed towards job creation - its geography and its social and educational systems are all more compatible with ecological goals than capitalism.

Ecocentric thinkers and activists should combine with and encourage the trades union and labour movement to work for the success of these alternatives to capitalism. For their growth at present lacks coordination and direction; they need encouragement and a sense of solidarity with each other - and they may need money for the establishment of an infrastructure. They are beginning to be so encouraged - for example, the South-West region of the Transport and General Workers' Union has launched a trustee scheme which will produce funds of £3m that will be used to set up new workers' cooperatives. Hence the 'purchasing power of trade unionists is used to create industrial jobs in S-W Britain' (*Guardian* 9.12.83). Ecocentrics can help to ensure that the growth of such socialistic experiments will avoid the pitfalls of centralisation, bureaucracy and a crude materialist outlook which have so bedevilled other similar developments. This difficult task will be made all the more so by success. For eventually the alternative mode of production will pose a threat to capitalism, because in by-passing it it will provide a far more acceptable society for ordinary people, who will 'vote with their feet'. By then, however, very many of us, not just a tiny educated elite, may have acquired heightened social and ecological consciousness and a sense of self-determination. We shall be less vulnerable to the fatalistic message that 'there is no alternative' to the disastrous policies which currently bring benefit to such a small minority of the world's people. And it is to be hoped that true critical education will play its part in widening this kind of consciousness, nourished by 'the economics of the epoch'. Set in this context, education will then proclaim to people loudly and clearly the kind of message which Vince Taylor (1980) gives his science student - a message which, it is hoped, has also been conveyed by this book:

> You will not experience the arbitrary nature of your beliefs by reading more scientific, analytical books, or by just thinking about them. Something or someone outside of you must jolt you into opening your eyes, perhaps just for a moment, to an aspect of reality that doesn't fit comfortably into your present belief structure. If this happens, hang on to it! Expand on it, explore it. Don't suppress and deny it. Rather ask whether some of your previously-held beliefs need to be opened up to make room for a richer reality.

REFERENCES

Abbey, E. (1975) *The Monkey Wrench Gang*, New York: Avon Books
Agricultural Advisory Council (1970) *Modern Farming and the Soil*, London: HMSO
Albury, D. and Schwartz, J. (1982) *Partial Progress: the Politics of Science and Technology*, London: Pluto Press
Allaby, M. (1970) 'One jump ahead of Malthus: can we avoid a world famine?', *Ecologist*, 1(1), 24-28
Allaby, M. (1971) *The Eco-Activists*, London: Charles Knight
Allaby, M. (1980) 'Malthus reinterred', *The Ecologist*, 10(6/7), Jul/Aug/Sept, 195-199
Allsopp, B. (1972) *Ecological Morality*, London: Muller
Apple, M. (1979) *Ideology and Curriculum*, London: Routledge and Kegan Paul
Apple, M. (1982) *Education and Power*, London: Routledge and Kegan Paul
Attfield, R. (1983) 'Christian attitudes to nature', *Journal of the History of Ideas*, 44(3), 369-386
Bach, R. (1977) *Illusions: the Adventures of a Reluctant Messiah*, London: Pan
Bahro, R. (1982A) 'A new approach for the peace movement in Germany', in Thompson, E.P. (ed) *Exterminism and Cold War*, London: New Left Books, 87-117
Bahro, R. (1982B) *Socialism and Survival*, London: Heretic Books
Barratt Brown, M. (1976) 'The crisis of capitalism and community production', in Barratt Brown *et al* (eds) *Resources and the Environment: a Socialist Perspective*, Nottingham: Spokesman Books, 5-9
Barrows, H. (1923) 'Geography as human ecology', *Annals, Association of American Geographers*, 13, 1-14
Bellamy, D. (1983) 'We need a Bill of Rights for nature', *FoE Newspaper*, Autumn, 21
Benn, T. (1980) 'The mandarins in Britain work as closely with the Emperor or Empress as can ever have occurred in ancient China', *The Guardian*, February 4
Bertalanffy, L. von (1968) *General System Theory: Foundations, Development, Applications*, New York: George Braziller
Blowers, A. (1984) 'The triumph of material interests - geography, pollution and the environment', *Political Geography Quarterly*, 3(1), 49-68
Blythe, R. (1969) *Akenfield*, Harmondsworth: Penguin
Bookchin, M. (1979) 'Ecology and revolutionary thought', *Antipode*, 10(3)/11(1), 21-32
Bookchin, M. (1980) *Towards an Ecological Society*, Montreal: Black Rose Books
Bottomore, T. (1983) 'Sociology', in McLellan, D. (ed), *Marx: the First 100 Years*, Oxford: OUP/Fontana, 103-142
Bowles, S. and Gintis, H. (1976) *Schooling in Capitalist America*, London: Routledge and Kegan Paul
Brand, S. (ed) (1971) *The Whole Earth Catalog*, Menlo Park, Calif.: Portola Institute
Bregman, S. (1982) 'Uranium mining on Indian lands', *Environment*, 24(7), 6-13, 33
Breitbart, M. (1979) Introduction to the special issue on anarchism and the environment, *Antipode*, 10(3)/11(1), 1-5
BSSRS (British Society for Social Responsibility in Science) (1982) *Science on OUR Side: a New Socialist Agenda for Science, Technology and Medicine*, London: BSSRS

Buchanan, C. (1981) *No Way to the Airport*, London: Longmans
Buchanan, K. (1973) 'The white north and the population explosion', *Antipode*, 5(3), 7-15
Bullock, A. and Stallybrass, O. (eds) (1977) *The Fontana Dictionary of Modern Thought*, London: Fontana
Bunge, W. (1973) 'The Geography', *Professional Geographer*, XXV, 331-7
Burch, G. and Pendell, E. (1947) *Human Breeding and Survival: Population Roads to Peace or War*, New York: Penguin
Burgess, R. (1978) 'The concept of nature in geography and Marxism', *Antipode*, 10(2), 1-11
Burton, I., Kates, R. and White, G. (1978) *The Environment as Hazard*, New York: OUP
Buttimer, A. (1976) 'Grasping the dynamism of lifeworld', *Annals, Association of American Geographers*, 66(2), 277-92
Bynum, W. (1975) 'The Great Chain of Being after forty years: an appraisal', *History of Science*, xiii, 1-28
Callenbach, E. (1978) *Ecotopia*, London: Pluto Press
Campbell, A., Keen, C., Norman, G. and Oakeshott, R. (1977) *Worker-Owners: the Mondragon Achievement*, London: Anglo-German Foundation for the Study of Industrial Society·
Capra, F. (1975) *The Tao of Physics*, London: Fontana
Capra, F. (1982) *The Turning Point*, London: Wildwood House
Castells, M. (1978) *City, Class and Power*, London: Macmillan
Chase, A. (1980) *The Legacy of Malthus: the Social Costs of the New Scientific Racism*, Urbana: Univ. Illinois Press
Chorley, R. (1973) 'Geography as human ecology', in Chorley, R. (ed) *Directions in Geography*, London: Methuen, 155-70
Coates, K. (1976) 'Needs', in Barratt Brown, M., Emerson, T. and Stoneman, C. (eds) *Resources and the Environment*, Nottingham: Spokesman Books
Cobbett, W. (1830) *Rural Rides*, Harmondsworth: Penguin
Cole, H., Freeman, C., Jahoda, M. and Pavitt, K. (1973) *Thinking About the Future: a Critique of the Limits to Growth*, London: Chatto and Windus, for Sussex Univ. Press
Commoner, B. (1967) *Science and Survival*, New York: Compass Books
Cosgrove, D. (1979) 'Ron Johnston and structuralism', *Journal of Geography in Higher Education*, 3(1), 107-111
Cotgrove, S. (1975) 'Technology, rationality and domination', *Social Studies of Science*, 5, 55-78
Cotgrove, S. (1982) *Catastrophe or Cornucopia: the Environment, Politics and the Future*, Chichester: Wiley
Cotgrove, S. and Duff, A. (1980) 'Environmentalism, middle class radicalism and politics', *Sociology Review*, 28, 335-51
Cottingham, J. (1978) '"A brute to the brutes?" Descartes' treatment of animals', *Philosophy*, 53, 551-59
Cranston, M. (1966) Introduction to his translation of Rousseau, Jean-Jacques (1743) *The Social Contract*, Harmondsworth: Penguin
Crosland, A. (1971) *A Social Democratic Britain*, Fabian Tract No. 404
Curtis, L., Courtney, F. and Trudgill, S. (1976) *Soils in the British Isles*, Ch. 15 'Some problems of modern farming', London: Longmans
Darwin, C. (1885) *The Origin of Species*, London: Murray (6th ed), first published 1859
Davies, P. (1983) *God and the New Physics*, London: Dent
Del Sesto, S. (1980) 'Conflicting ideologies of nuclear power: Congressional testimony on nuclear reactor safety', *Public Policy*, 28(1), 39-70

Doughty, R. (1981) 'Environmental theology: trends and prospects in Christian thought', *Progress in Human Geography*, 5(2), 234-248

The Ecologist, (1972) 'Blueprint for survival', *The Ecologist*, 2(1), 1-43

Ecology Action East (1970) 'The power to destroy, the power to create', in Disch, R. (ed) *The Ecological Conscience: Values for Survival*, N. Jersey: Prentice Hall

Ecology Party (1979) *The Real Alternative*, election manifesto, Birmingham: The Ecology Party

Ecology Party (1983) *Politics for Life*, election manifesto, London: The Ecology Party

Economist (1981) 'The nature of knowledge', December 26

Edgley, R. (1983) 'Philosophy', in McLellan, D. (ed), *Marx: the First 100 Years*, Oxford: OUP/Fontana, 239-302

Edwards, P. (ed)(1972) *The Encyclopaedia of Philosophy*, vol 7, 206-9, London: Collier-Macmillan

Ehrlich, P. (1969) 'Eco-catastrophe', *Ramparts*, 8(3), 24-28

Ehrlich, P. (1970) *The Population Bomb*, New York: Ballantine Books

Ehrlich, A. and Ehrlich, P. (1972) *Population, Resources, Environment*, San Francisco: Freeman

Erhlich, P. and 19 other authors (1983) 'Long-term biological consequences of nuclear war', *Science*, 222(4630), 23 Dec., 1293-1300

Elliot, D. (1977) *The Lucas Aerospace Workers' Campaign*, Young Fabian Pamphlet 46, London: Fabian Society

Emerson, T. (1976A) 'An indictment of capitalism', in Barratt Brown, M. *et al* (eds) *Resources and the Environment*, Nottingham: Spokesman Books, 21-25

Emerson, T. (1976B) 'Flawed economics, flawed technology', in Barratt Brown, M. *et al* (eds) *Resources and the Environment*, Nottingham: Spokesman Books, 64-75

Encyclopaedia Britannica (1978) 'Existentialism', *Macropaedia*, 7, 73-79, Chicago: Encyclopaedia Britannica

Engels, F. (1963) *Dialectics of Nature*, Moscow: Foreign Languages Publishing (First published 1890)

Ensenberger, H. (1974) 'A critique of political ecology', *New Left Review*, 84, 3-32

Environmental Fund (1977) *Behind the 'Food Crisis'*, Washington DC: The Environmental Fund

Erisman, F. (1973) 'The environmental crisis and present-day romanticism: the persistence of an idea', *Rocky Mountain Social Science Journal*, 10, 7-14

Evans, H. (1983) *Good Times: Bad Times*, London: Weidenfield and Nicholson

Ezrahi, Y. (1971) 'The political resources of American science', *Science Studies*, 1, 121

Fairhall, J. (1983) 'A bizarre vision of social engineering', *The Guardian*, 22 February

Febvre, L. (1924) *A Geographical Introduction to History*, London: Routledge and Kegan Paul (reissued 1966)

Fowles, J. (1982) 'The Falklands and a death foretold', *The Guardian*, August 14

Fraser-Darling, F. (1971) *Wilderness and Plenty*, New York: Ballantine

Freeman, C. (1973) 'Malthus with a computer', in Pavitt *et al*, *Thinking About the Future*, London: Chatto and Windus, 5-13

Friere, P. (1972) *Pedagogy of the Oppressed*, Harmondsworth: Penguin

Fry, C. (1975) 'Marxism and ecology', *The Ecologist*, 6(9), 328-332

Galbraith, J. (1958) *The Affluent Society*, Harmondsworth: Penguin

Giddings, R. (1978) 'A myth riding by', *New Society*, 46, 588-9

Glacken, C. (1967) *Traces on the Rhodian Shore*, Berkeley: Univ. Calif. Press

Gold, J. (1980) *Introduction to Behavioural Geography*, Oxford: OUP

Goldsmith, E. (1975A) 'Is science a religion?', *The Ecologist*, 5(2), 50-62

Goldsmith, E. (1975B) 'The fall of the Roman Empire', *The Ecologist*, 5(6), 196-206

Goldsmith, E. (1978) 'The religion of a stable society', *Man-Environment Systems*, 8, 13-24

Golub, R. and Townsend, J. (1977) 'Malthus, multinationals and the Club of Rome', *Social Studies of Science*, 7, 202-222

Gorz, A. (1976) 'On the class character of science and scientists', in Rose, H. and Rose, S. (eds), *The Political Economy of Science*, London: Macmillan, 59-71

Gorz, A. (1980) *Ecology as Politics*, Boston, Mass: South End Press

Granada TV (1983) *The Power Brokers*, 'World in Action' series, transmitted in January

Hales, M. (1982) *Science or Society?*, London: Pan Books/Channel 4

Hall, P. (1981) *Great Planning Disasters*, Harmondsworth: Penguin

Hamer, M. (1974) *Wheels within Wheels*, London: Friends of the Earth

Hardin, G. (1949) *Biology: Its Human Implications*, San Francisco: W.H. Freeman

Hardin, G. (1964) *Population, Evolution and Birth Control: a Collage of Controversial Ideas*, San Francisco: Freeman

Hardin, G. (1968) 'Tragedy of the commons', *Science*, 162, 1243-1248

Hardin, G. (1974) 'Living on a lifeboat', *BioScience*, 24, 10

Hardy, D. (1979) *Alternative Communities in Nineteenth-Century England*, London: Longman

Harris, K. (1979) *Education and Knowledge*, London: Routledge and Kegan Paul

Harrison, F. (1982A) *Strange Land: the Countryside, Myth and Reality*, London: Sidgwick and Jackson

Harrison, H. (1982B) *Make Room! Make Room!*, Harmondsworth: Penguin (2nd ed), filmed as 'Soylent Green', first published 1966

Harvey, D. (1973) *Social Justice and the City*, London: E. Arnold

Harvey, D. (1974A) 'What kind of geography for what kind of public policy?', *Transactions, Institute of British Geographers*, 63, 18-24

Harvey, D. (1974B) 'Population, resources and the ideology of science', *Economic Geography*, 50, 256-277

Harvey, D. (1982) *The Limits to Capital*, Oxford: Blackwells

Hollingsworth, T. (1973) Introduction to Malthus's *Essay on the Principle of Population*, 7th ed. 1872 text, London: Dent

Holton, G. (1956) 'Johannes Kepler's universe: its physics and metaphysics', *American Journal of Physics*, 24, 340-351

Hoskins, W. (1955) *The Making of the English Landscape*, London: Hodder and Stoughton

Howard, E. (1902) *Garden Cities of Tomorrow*, London: Faber and Faber 1974 ed

Huanxuan, G. (1981) 'Environmental protection in China', *Beijing Review*, 26, 12-15

Huckle, J. (ed) (1983) *Geographical Education: Reflection and Action*, Oxford: OUP

Hull, D. (1973) *Darwin and his Critics*, Harvard UP

Huntington, E. (1907) *The Pulse of Asia*, Boston: Houghton Mifflin

Huntington, E. (1915) *Civilisation and Climate*, Yale UP

Huntington, E. (1945) *Mainsprings of Civilisation*, New York: Wiley

IFR (International Fellowship of Reconciliation) (1972) *The Menton Statement*, New York: IFR

James, P.E. (1972) *All Possible Worlds: a History of Geographical Ideas*, Indianapolis: The Odyssey Press

James, P. (1979) *Population Malthus: his Life and Times*, London: Routledge and Kegan Paul

Jeans, D. (1974) 'Changing formulations of the man-environment relationship in Anglo-American geography', *Journal of Geography*, 73(3), 36-40

Johnston, R. (1979) *Geography and Geographers*, London: E. Arnold

Johnston, R. (ed) (1981) *The Dictionary of Human Geography*, Oxford: Blackwells

Johnston, R. (1983) *Philosophy and Human Geography*, London: E. Arnold

Jungk, R. (1982) *Brighter Than a Thousand Suns*, Harmondsworth: Penguin, originally published 1956

Kahn, H. (1980) 'Facing the future', from Kahn, H. and Ford, T. *The Optimists*, Farmington, Conn.: Emhart Corporation

Kahn, H. and Wiener, A. (1968) 'Man's Faustian powers' from Kahn, H. and Wiener, A. *Environment and Change: the Next Fifty Years*, Bloomington, Ind.: Indiana Univ. Press

Kiernan, V. (1983) 'History', in McLellan, D. (ed) *Marx: the First 100 Years*, Oxford: OUP/Fontana, 57-102

Kimber, R. and Richardson J. (eds) (1974) *Campaigning for the Environment*, London: Routledge and Kegan Paul

Kirk, W. (1963) 'Problems of geography', *Geography*, 48, 357-71

Koestler, A. (1964) *The Sleepwalkers*, Harmondsworth: Penguin

Kormondy, E. (1969) *Concepts of Ecology*, N. Jersey: Prentice Hall

Kropotkin, P. (1899) *Fields, Factories and Workshops (Tomorrow)*, ed by Colin Ward, 1974, London: Unwin

Kuhn, T. (1962) *The Structure of Scientific Revolutions*, Chicago: Univ. Chicago Press

Laszlo, E. (1983) *Systems Science and World Order*, Oxford: Pergamon Press

Learmonth, A. and Simmons, I. (1977) 'Man Environment Relationships as Complex Ecosystems', Unit 8 of *Fundamentals of Human Geography*, D204, Milton Keynes: Open University

Leopold, A. (1949) *A Sand County Almanack*, New York: OUP

Ley, D. (1978) 'Social geography and social action', in Ley, D. and Samuels, M. (eds) *Humanistic Geography: Prospects and Problems*, London: Croom Helm, 41-57

London, H. (1969) 'American romantics: old and new', *Colorado Quarterly*, 8, 5-20

Lovejoy, A. (1974) *The Great Chain of Being*, Cambridge, Mass: Harvard UP. Twelfth printing.

Lovelock, J. (1979) *Gaia*, New York: OUP

Lowe, P. and Goyder, J. (1983) *Environmental Groups in Politics*, London: George Allen and Unwin

Lowe, P. and Warboys, M. (1978) 'Ecology and the end of ideology', *Antipode*, 10(2), 12-21

LWT (London Weekend Television) (1983) *Sizewell and the Nuclear Scientists*, 'TV Eye' programme, transmitted 16 September

McFadden, C. (1977) *The Serial: a Year in the Life of Marin County*, New York: Knopf

McHarg, I. (1969) *Design with Nature*, New York: Natural History Press

McLellan, D. (1975) *Marx*, Glasgow: Collins (Fontana Modern Masters)

McRobie, G. (1982) *Small is Possible*, London: Abacus Books

Maddox, J. (1972) *The Domesday Syndrome*, London: Macmillan

Malthus, T. (1872) *An Essay on the Principle of Population*, 7th ed., London: Dent

Marx, K. (1959) *Capital Vol. 1*, Moscow: Foreign Languages Publishing House, English ed. first published 1887

Marx, K. (1981) *Grundrisse*, Harmondsworth: Penguin, first published 1939

Marx, L. (1973) 'Pastoral ideals and city troubles', in Barbour, (ed) *Western Man and Environmental Ethics*, London: Addison Wesley, 93-115

Matley, I. (1966) 'The Marxist approach to the geographical environment', *Annals, Association of American Geographers*, 56, 97-111

Matley, I. (1982) 'Nature and society: the continuing Soviet debate', *Progress in Human Geography*, 6(3), 367-396

Maunder, W. (1970) *The Value of the Weather*, London: Methuen

Meacher, M. (1976) 'Global resources, growth and political agency', in Barratt Brown, M., Emerson, C. and Stoneman, C. (eds) *Resources and the Environment: a Socialist Perspective*, Nottingham: Spokesman Books, 42-47

Meadows, D., Meadows, D., Randers, J., Behrens, W. (1972) *The Limits to Growth*, London: Earth Island

Melchett, P. (1981) 'Bill made Act', *Ecos*, 2(4)

Merton, R. (1972) 'The institutional imperatives of science', in Barnes, B. (ed) *Sociology of Science*, Harmondsworth: Penguin

Mills, W. (1982) 'Metaphorical vision: changes in Western attitudes to the environment', *Annals, Association of American Geographers*, 72(2), 237-253

Moss, R. (1979) 'On geography as science', *Geoforum*, 10, 223-233

Muir, J. (1898) 'The wild parks and forest reservations of the West', *Atlantic Monthly*, LXXXI, 483

Nash, R. (1974) *Wilderness and the American Mind*, New Haven, Conn.: Yale UP

Nelkin, D. (1975) 'The political impact of technical expertise', *Social Studies of Science*, 5, 35-54

Neuhaus, R. (1971) *In Defense of People: Ecology and the Seduction of Radicalism*, New York: Macmillan

Newby, H. (1979) *Green and Pleasant Land?*, Harmondsworth: Penguin

Nicholson, M. (1970) *The Environmental Revolution*, Harmondsworth: Penguin

Norwine, J. (1981) 'Geography as human ecology? The man-environment equation reappraised', *International Journal of Environmental Studies*, 17, 179-190

O'Flinn, P. (1978) 'Freedom for a few', review of Callenbach's '*Ecotopia*', *Socialist Review*, September

Opie, J. (ed) (1971) *Americans and the Environment: the Controversy over Ecology*, Lexington, Mass.: D.C. Heath

O'Riordan, T. (1977) 'Environmental ideologies', *Environment and Planning, Series A*, 9, 3-14

O'Riordan, T. (1981A) *Environmentalism*, London: Pion

O'Riordan, T. (1981B) 'Environmentalism and education', *Journal of Geography in Higher Education*, 5(1), 3-18

Orwell, G. (1981) *1984*, Harmondsworth: Penguin, originally published 1948

Packard, V. (1963) *The Waste Makers*, Harmondsworth: Pelican

Passmore, J. (1974) *Man's Responsibility to Nature*, London: Duckworth

Pavitt, K. (1973) 'Malthus and other economists', in Cole *et al*, *Thinking About the Future*, London: Chatto and Windus, 137-158

Pepper, D. (1980) 'Environmentalism, the "lifeboat ethic" and anti-airport protest', *Area*, 12(3), 177-82

Pepper, D. (1983) 'Bringing physical and human geographers together: why is it so difficult?', in Cannon, T., Forbes, M. and Mackie, J. (eds) *Society and Nature*, London, Union of Socialist Geographers, 19-31

Perelman, M. (1979) 'Marx, Malthus and the concept of natural resource scarcity', *Antipode*, 11(2), 80-89

Perkins, J. (1984) 'A case of industrial pollution in 19th-century Rouen', in preparation

Petersen, W. (1979) *Malthus*, London: Heinemann

Peterson, J. (ed) (1983) *Nuclear War: the Aftermath*, Oxford: Pergamon (based on a special edition of *Ambio*)

Philo, G., Hewitt, J., Beharrell, P. Davis, H. (1982) *Really Bad News*, London: Writers and Readers Publishing Co-operative

Pinchot, G. (1910) *The Fight for Conservation*, New York: Harcourt Brace

Pirsig, R. (1974) *Zen and the Art of Motor Cycle Maintenance*, London: Transworld Publishers (Corgi)

Pringle, P. and Spigelman, J. (1982) *The Nuclear Barons*, London: M. Joseph

Prior, M. (1954) 'Bacon's man of science', *Journal of the History of Ideas*, XV, 41-54

Pryce, R. (1977) 'Approaches to the study of man and the environment', Unit 2 of Section 1 of *Fundamentals of Human Geography*, Milton Keynes: Open University course D204

Quaini, M. (1982) *Geography and Marxism*, Oxford: Blackwells, translated by A. Braley

Ravetz, J. (1971) *Scientific Knowledge and its Social Problems*, Oxford: Clarendon Press

Rees, J. (1983) 'Reaping the profit of greed', *The Guardian*, 23 June

Reich, C. (1970) *The Greening of America*, New York: Random House

Richards, S. (1983) *Philosophy and Sociology of Science: An Introduction*, Oxford: Blackwells

Robins, K. and Webster, F. (1983) 'The mis-information society', *New Universities Quarterly*, 37(4), 344-55

Roszak, K. (1970) *The Making of a Counter Culture*, London: Faber

Rowles, G. (1978) 'Reflections on experiential fieldwork', in Ley, D. and Samuels, M. (eds) *Humanistic Geography*, London: Croom Helm, 173-93

Roxborough, I. (1983) 'The friendly approach', *The Guardian*, 21 June

Russell, B. (1946) *History of Western Philosophy*, London: Unwin (1979 ed)

Russwurm, L. (1974) 'A systems approach to the natural environment', in Russwurm, L. and Sommerville, E. *Man's Natural Environment: a Systems Approach*, Massachusetts: Duxbury Press

Ryabchikov, A. (1976) 'Problems of the environment in a global aspect', *Geoforum*, 7, 107-113

Ryder, N. (1983) *Science, Television and the Adolescent*, London: Independent Broadcasting Authority

Sandbach, F. (1978A) 'A further look at the environment as a political issue', *International Journal of Environmental Studies* 12, 99-110

Sandbach, F. (1978B) 'Ecology and the "Limits to Growth" debate', *Antipode*, 10(2), 22-32

Sandbach, F. (1980) *Environment, Ideology and Policy*, Oxford: Blackwell

Santmire, P. (1973) 'Historical dimensions of the American crises', in Barbour, I. (ed) *Western Man and Environmental Ethics*, London: Addison-Wesley, 66-92

Sartre, J.P. (1943) *Being and Nothingness*, translated by H. Barnes, 1957, London: Methuen

Sartre, J.P. (1946) *Existentialism and Humanism*, translated by P. Mairet, 1948, London: Methuen

Sayer, A. (1979) 'Epistemology and conceptions of people and nature in geography', *Geoforum*, 10, 19-43

Schell, J. (1982) *The Fate of the Earth*, Harmondsworth: Penguin

Schmidt, A. (1971) *The Concept of Nature in Marx*, London: New Left Books

Schnaiberg, A. (1980) *The Environment: From Surplus to Scarcity*, New York: Oxford University Press

Schumacher, F. (1973) *Small is Beautiful: Economics as if People Really Mattered*, London: Abacus

Schumacher, F. (1980) *Good Work*, London: Abacus

Seamon, D. (1979A) *A Geography of the Lifeworld*, London: Croom Helm

Seamon, D. (1979B) 'Phenomenology, geography and geographical education', *Journal of Geography in Higher Education*, 3(2), 40-50

Semple, E. (1911) *Influences of Geographical Environment*, New York: Henry Holt

SERA (Socialist Environment and Resources Association) (1980) *Ecosocialism in a Nutshell*, London: Writers and Readers Publishing Cooperative

Shabad, T. (1979) 'Communist environmentalism', *Problems of Communism*, May-June, 64-67

Shoard, M. (1980) *The Theft of the Countryside*, London: Temple Smith

Shoard, M. (1982) 'The lure of the moors', in Gold, J. and Burgess, J. (eds) *Valued Environments*, London: George Allen and Unwin, 55-73

Simon, J. (1981) *The Ultimate Resource*, Oxford: Martin Robertson

Skinner, B. (1948) *Walden II*, New York: Macmillan, reprinted 1976

Skolimowski, H. (1981) *Eco-Philosophy*, London: Marion Boyars

Smith, N. and O'Keefe, P. (1980) 'Geography, Marx and the concept of nature', *Antipode*, 12(2),30-39

Spoehr, A. (1967) 'Cultural differences in the interpretation of natural resources', in Thomas, W. (ed) *Man's Role in Changing the Face of the Earth*, 93-101

Stephens, M. (1980) *Three Mile Island: the Hour by Hour Account of What Really Happened*, London: Junction Books

Stretton, H. (1976) *Capitalism, Socialism and the Environment*, Cambridge: Cambridge University Press

Stoddart, D. (1965) 'Geography and the ecological approach: the ecosystem as a geographic principle and method', *Geography*, 50, 242-51

Stoddart, D. (1966) 'Darwin's impact on geography', *Annals, Association of American Geographers*, 56, 683-98

Stuart, A. (1958) *Overpopulation: Twentieth Century Nemesis*, New York: Exposition Press

Tatham, G. (1951) 'Environmentalism and possibilism', in Taylor, G. (ed) *Geography in the Twentieth Century*, London: Philosophical Library, 128-162

Taylor, G. (1937) *Environment, Race and Migration*, Toronto: University of Toronto Press

Taylor, G. (1940) *Australia: a Study of Warm Environments and Their Effect on British Settlement*, London: Methuen

Taylor, C. (1982) 'Marxist philosophy', in Magee, B. *Men of Ideas*, Oxford: OUP

Taylor, V. (1980) 'Subjectivity and science: a correspondence about belief', *The Ecologist*, 10(6-7), 230-234

TGWU (Transport and General Workers' Union) (1983) *A Better Future for Defence Jobs*, London: TGWU

Thackray, A. (1974) 'Natural knowledge in a cultural context: the Manchester model', *American Historical Review*, 79, 672-709

Thomas, K. (1983) *Man and the Natural World: Changing Attitudes in England 1500-1800*, London: Allen Lane

Thoreau, H. (1974) *Walden*, New York: Collier Books (8th printing)

Thyme, J. (1978) *Motorways versus Democracy*, London: Macmillan

Transport 2000 (1983) *A Reproduction of the Letter and Memorandum About a Proposed Inquiry into Heavier Lorries from Mr Joseph Peeler to Mr Peter Lazarus*, Document 419, presented to the M40 Inquiry, Banbury: Oxon

Tuan Yi Fu (1968) 'Discrepancies between environmental attitude and behaviour: examples from Europe and China', *Canadian Geographer*, 12, 176-91

Tuan Yi Fu (1971A) 'Geography, phenomenology and the study of human nature', *Canadian Geographer*, 15(3), 181-192

Tuan Yi Fu (1971B) *Man and Nature*, Resource Paper No 10, Washington DC: Association of American Geographers

Tuan Yi Fu (1972) 'Structuralism, existentialism and environmental perception', *Environment and Behaviour*, 4(3), 319-331

Tuan Yi Fu (1974) *Topophilia: a Study of Environmental Perception, Attitudes and Values*, N. Jersey: Prentice Hall

Tucker, W. (1981) 'Some second thoughts on environmentalism', *Dialogue*, 53(3), 31-34

Turco, R., Toon, O., Ackerman, T., Pollock, J., Sagan, C. (1983) 'Nuclear winter: long-term consequences of multiple nuclear explosions', *Science*, 222(4630), 23 December, 1283-1292

US Interagency Committee (1982) *The Global 2000 Report to the President*, Harmondsworth: Penguin Books (First Publ. US 1980)

Voght, J. (1948) *Roads to Survival*, New York: Wm. Sloane Associates

Walker, M. (1983) 'Tory historians find a heritage to nationalise', *Guardian*, 21 March

Walmsley, D. (1974) 'Positivism and phenomenology in human geography', *Canadian Geographer*, 18, 95-106

Warnock, M. (1979) *Existentialism*, Oxford: OUP

Webster, F. and Robins, K. (1982) 'New technology: a survey of trade union response in Britain', *Journal of Industrial Relations*, 13(1), 7-26

Wheeler, J. (1973) 'From relativity to mutability', in Mehra, J. (ed) *The Physicist's Conception of Nature*, Dordrecht: D. Reidel, 202-247 (p244)

White, G. (1965) 'Contributions of geographical analysis to river basin development', in Burton, I. and Kates, R. (eds) *Readings in Resource Management and Conservation*, Chicago: University Press, 375-394

White, G. (1973) 'Natural hazards research', in Chorley, R. (ed) *Directions in Geography*, London: Methuen, 193-216

White, L. (1967) 'The historical roots of our ecologic crisis', *Science*, 155, 1203-07

Williams, R. (1975) *The Country and the City*, St Albans: Paladin

Williams, R. (1983A) *Keywords*, London: Fontana (Flamingo) revised

Williams, R. (1983B) 'Culture', in McLellan, D. (ed) *Marx: the First 100 Years*, Oxford: OUP/Fontana, 15-56

Zaring, J. (1977) 'The romantic face of Wales', *Annals, Association of American Geographers*, 67(3), 397-418

APPENDIX: A GLOSSARY OF SOME TERMS USED IN THIS BOOK

This glossary draws particularly on Bullock and Stallybrass - 'Fontana Dictionary of Modern Thought'; Richards - 'Philosophy and Sociology of Science: an Introduction'; Williams - 'Keywords'; and Johnston - 'Philosophy and Human Geography'. It does not attempt to make all the definitions complete and rounded, but concentrates on those aspects of the definitions which are particularly relevant to this book.

In addition, readers are referred to Edwards, P. the 'Encyclopaedia of Philosophy', and Gould, J. and Kolb, W. 'A Dictionary of the Social Sciences'.

A-HISTORICAL: see Historical

ALIENATION A complicated term, because it is used in many specific technical contexts. In ecocentrism it is used in a general sense, to indicate man's supposed loss of his 'original' human nature and characteristics through the development of an over-sophisticated, artificial society. It includes, then, the idea of estrangement of humans from both their own (human) and external nature, and from each other. This comes through feelings of inability to affect the character and development of society, and through the dehumanisation which stems from a large-scale, bureaucratic, centralised and secretive society in which mechanisation and the division of labour have invaded all walks of life. In Marxism, the term is used in analysing class society and capitalist relations of production, where the worker loses control over the production process and over what he produces, and he becomes a commodity, or thing, related to other people and to nature only through the cash nexus.

ANALYSIS: see Reductionism and Holism.

ANARCHISM A political movement, largely on the left, which opposes centralised state control and 'representative democracy'. Instead it advocates self-government, organised in non-hierarchical small collectives (based, in anarcho-syndicalism, on trades unions), where everyone can participate in democratically-made decisions, and where any law is obeyed only because people want to obey it (i.e. it represents the genuine collective wish of the people). Anarchy, then, would abolish the state and replace it by free association and voluntary cooperation of individuals and groups (e.g. in federations). Anarchy emphasises the importance of the individual, and, in the view of many, communism submerges the individual - ANARCHO-COMMUNISM reconciles these two opposites. It holds that individual freedom and fulfilment are fully realised only through relating to other people and the community as a whole.

ANARCHO-COMMUNISM: see Anarchism.

BIOETHIC The notion, in romanticism and ecocentrism, of man's moral duty towards, and respect and reverence for, nature. It holds that the natural world has *biotic rights*, including the right to existence, which are quite independent of any considerations of its usefulness to man. Bioethical principles are invoked heavily in modern attempts to legislate for the protection of wildlife and wildernesses (O'Riordan 1981A).

BOURGEOIS: see Capitalism.

CAPITALISM In the Marxist sense, the two-class economic system whereby one class (BOURGEOISIE) owns the means of production, distribution and exchange, while the other has only its labour to sell (PROLETARIAT). By virtue of his monopoly, the former can exchange the goods produced with his resources at a value in excess of that representing his worker's remuneration - the labour value, or what is required for the subsistence of the workforce. This surplus value represents the profit, and the express purpose of the capitalist is to accumulate this profit - the 'capital'. It follows that the greater the difference between labour value and exchange value (a difference known technically as *exploitation*), the greater is the rate of capital accumulation. In the broader sense, capitalism is regarded as the system of private ownership and profit regulated by laissez-faire economics and ideology in the West.

CARTESIAN DUALISM: see Objective.

DETERMINISM The doctrine that human action is not free, but is determined, and therefore constrained, by forces external to human will (forces such as natural or economic or historical 'laws', or the law or design of a god). Laws are statements describing deterministic relationships, for they describe how the behaviour and characteristics of given (dependent) variables follow from the behaviour and characteristics of other (independent) variables, such that a change in the latter produces a necessary and predictable change in the former. It follows that if human action and behaviour is determined it can be described and predicted by law-like statements, and that humans are not totally creatures of free will, and therefore they are not totally responsible for what they do.

DIALECTIC In the basic sense, the art of investigating truth by logical disputation (i.e. through contradiction in discussion). Through this process of contradiction of ideas and statements - through the 'fruitful collision of ideas' - higher truth is reached. It is, then, a mode of investigation which differs from analysis, for knowledge is advanced by resolving and *synthesising* opposites. This notion, of the interaction of opposing or contradictory forces, was taken by German idealist philosophers (especially Hegel) to apply not just to the art of dispute, but to reality. Dialectical criticism was a way of making sense of contradictions in thought (and in world history, which Hegel imagined was the objective manifestation of thought) and of unifying them to discover a higher truth - about the nature of 'things-in-themselves'. But as an idealist Hegel made 'the spirit' (including ideas and metaphysics) primary, and the material world of secondary importance. *Marxism*, in its HISTORICAL MATERIALISM, reverses these priorities. It takes historical and political processes and events to be the outcome of conflict between opposing social forces (e.g. opposing classes) such opposition is a function of contradictions arising in the process of producing to meet material needs, and organising this production. This interaction of opposing social forces is the chief mechanism for social change and progress - for the creation of new modes of production on which new societies are based. In its more crude form, the idea was expressed in the DIALECTICAL MATERIALISM of Engels. This held that the dialectic was not only the underlying 'law of motion' of historical development, it also applied to the natural world. In fact nature, society and the process of thinking had in common one underlying general law - the dialectic. For many Marxists this is too sweeping a concept, but they would admit of a dialectical relationship between nature and society, in which the two constantly interact to change one another. The kind of relationship involved in the dialectic, then, is very much more complex than deterministic relationships, and the 'laws' involved in a dialectical conception of history are far less rigid than are those of deterministic classical science.

DIALECTICAL MATERIALISM: see Dialectic.

ECOCENTRISM A 'mode of thought' (O'Riordan) viewing man as a part of a global ecosystem, and subject to ecological and systems laws. These, and the demands of an ecologically-based morality, constrain human action, particularly through imposing limits to economic and population growth. There is also a strong sense of respect for nature in its own right, as well as for pragmatic 'systems' reasons. Ecocentrics lack faith in modern large-scale technology and the technical and bureaucratic elites, and they abhor centralisation and materialism. If politically to the right they may emphasise the idea of limits, advocating compulsory restraints on human breeding, levels of resource consumption and access to nature's 'commons'. If to the left, their emphasis may be more on decentralised, democratic, small-scale communities using 'soft' technology and renewable energy, 'acting locally and thinking globally'.

ECOLOGY That branch of biology which deals with the relationships between organisms and their environment (the latter including other organisms). Hence ECOSYSTEM: a functioning interactive system consisting of one or more living organisms and the physical, chemical and biological factors in their environment. All the elements of an ecosystem are interdependent, and they are linked by flows of energy and matter through them. THE ECOLOGY MOVEMENT is a very loosely-used term, that mainly seems to embrace people described here as 'ecocentrics' - those who derive morality and life-styles from ecological 'principles' (like carrying capacity; the idea of strength through diversity; of hierarchical organisation). Hence there is a strong tendency among them to equate human and natural societies.

ECOLOGY MOVEMENT: see Ecology.

ECOSYSTEM: see Ecology.

EMPIRICAL A term describing knowledge which is based on observation, experience and experiment - on, that is, the evidence of the *senses* (what can be seen, touched, smelt, heard). Or knowledge which can be *derived from* such evidence by methods of inductive logic, including mathematics and statistics. Hence, EMPIRICISM, the doctrine that knowledge *must*, in order to be valid, be derived from observed experience.

EMPIRICISM: see Empirical.

EXISTENTIALISM A philosophy which holds that reality is created by the free acts of humans - there are no 'external' determining factors, such as historical processes or natural laws and constraints. There are not even 'essences' to give meaning to phenomena (see phenomenology). We forge our individual selves out of 'such senseless circumstances' as are given us, and there is no single human essence. So our humanity and our values can be derived only from our own individual *human existence* - there are no other factors constraining our freedom to create our own world, except those which stem from other people doing the same thing (factors such as customs and social conventions). Thus free, we are also responsible for the world we have created, and whether it is good or bad. While we have to live in society we must behave as individuals, resisting the force of others trying to create a society which will determine and alienate us. This freedom of will, and consciousness of it, distinguishes and separates us from the rest of nature, which lacks such attributes.

FINAL CAUSE: see Physico-Theology.

HISTORICAL In the Marxist sense, to be historical is to put events, explanations, and 'laws' governing economic and social behaviour in their historical context as opposed to 'universalising' them. In particular this means seeking explanation in terms of the 'economics of the epoch', i.e. the way that ideas, historical events and social and intellectual movements are materially-based in specific modes and organisations of production (feudalism, capitalism, socialism) and the social relations that stem from them. Being A-HISTORICAL, therefore, involves explaining events in idealist terms, or as a result of the operation of 'laws' of nature and human behaviour (biologising history) or other forces which are supposedly not attributable to specific societies and cultures.

HISTORICAL MATERIALISM: see Dialectic.

HOLISM The doctrine (in opposition to reductionism) of the importance of the whole as more than just the sum of the parts. This leads, for example, to the investigation of complex systems as such - in terms of the relationships between the components and their functioning *together* - rather than through ANALYSING them (explaining the components separately and without relating them to the whole). Thus holism says that 'wholes' have characteristics that cannot be explained merely in terms of the properties of the constituents. Since this is true of organisms, then ORGANICISM, or the ORGANIC SYSTEMS view (Capra 1982) is a form of holism. This draws an analogy between complex systems and living organisms, where to examine separately or remove one part from the organism makes that part (and perhaps the whole system) meaningless.

IDEALISM Philosophical theory that there are no material things that exist independently of minds (e.g. the mind of God and/or the minds of men). Consequently idealists see social, economic and historical development as fashioned essentially from the introduction of new ideas, and the development of old ideas and values. Change can be seen, therefore, as the growth of reason. (See materialism, with which idealism is usually contrasted.)

IDEOLOGY As used in Marxism the term denotes a set of ideas, ideals, beliefs or values derived not from distinterested thought, but stemming from material vested interests. They are presented as being universal, but reflect those more limited interests. As such, an ideology supports an already-reached position, thus it is likely to contain bias, prejudice and unexamined assumptions about the world. Marxism sees an ideology as a set of justifications masking specific interests and perhaps constituting a formula whereby social action may be mobilised. In sociology, the concept of ideology may be used to describe a complete commitment to a particular way of life, and/or a complete interpretative scheme by which the world may be rendered more comprehensible - a world view.

LEGITIMATION Justification. Making acceptable - especially where one group exercises power over another group in the name of a principle jointly accepted as being universally true or beneficial.

LOGICAL POSITIVISM: see Positivism.

MATERIALISM Strictly speaking, holds that everything which exists is material, occupying some space at some time. Thus it denies substantial existence to minds as abstract entities (rather than as a function of the organism of the brain). But in terms of explanations of society and history, materialist analyses (for example, Marxism) see change and development stemming principally from material factors - such as the way we organise ourselves and our labour to gain our material subsistence through production. Materialism holds that the predominant ideas which shape social change are not autonomously and independently derived from abstract thinking and reasoning processes; rather they relate strongly to the mode and organisation of economic activity in society, and to concrete events which have already occurred in history.

OBJECTIVE A quality in objects which is independent of the subjects who examine them - a 'primary' quality, such as shape, size, position, motion, as opposed to a 'secondary' quality, such as smell, colour, goodness, sweetness. The primary qualities, it is said, will appear the same to everyone; secondary qualities will vary according to the disposition of the observer (the subject), because they are projections of the human mind - they have no existence independent of the observer. To accept that OBJECTIVITY, where the subject maintains detachment from the object he investigates, is possible, one must first accept Descartes' proposition that subject and object, mind and matter *can* be separated (THE CARTESIAM DUALISM). The term 'objective' is often used more loosely to mean 'unemotional', 'value-free' and/or not not biassed to any particular side (i.e. 'balanced'). Many scientists now believe that objectivity, in the strict sense, is not possible, and that the concept has use only as a part of a procedural method

ORGANICISM, ORGANIC SYSTEMS VIEW: see Holism.

PARADIGM Example, pattern. When related to a scientific discipline, (Kuhn), the word means the general set of rules and procedures under which people study and investigate the subject matter of the discipline. The paradigm for a discipline is also that definition, which most of its practitioners will agree on, of what are the main problems it tries to solve - it represents the consensus from within the subject on what the subject is about. From time to time this consensus may change, in response to developments in or outside the discipline. So progress may be by paradigm overturning - or 'revolution'. But it is usually difficult to gain acceptance for theories and research programmes which challenge ruling paradigms, e.g. creationism in biology, the diluvial theory in geology, or even the 'big bang' theory in astrophysics. Once the paradigm has been established, most research tends to support, substantiate and illuminate it; hence most work 'solves puzzles' set by the paradigm (and those people who developed it).

PHENOMENOLOGY A method of philosophical enquiry whose object is to describe and understand phenomena as experienced through human consciousness, without trying to explain them in the sense of deriving laws about what causes them to be as they are. Phenomenology holds that the world has no external, objective existence separate from man (and vice versa); rather, nature is structured through human consciousness. And since humans are unique individuals, generalisations about 'nature' are inappropriate - though we do often hold in common the identification of *essences* (essential characteristics or meanings) or phenomena. So-called objective 'facts' are insignificant compared with the inner experiences of individuals and groups of people. These condition how essences and meanings are attached to things via consciousness and human *intentions*, and how unspoken assumptions are made about the everyday world of familiar objects about us - our 'lifeworld'. Phenomenological investigations (as used, for example, in the social sciences) aim, among other things, to make such assumptions explicit. But first, the investigator must attempt to cast aside his own presuppositions about the world. Then his job will be to try to understand and describe (not explain) how others structure and perceive their lifeworld by developing a sympathetic understanding with them and how they live (through both living as they do, and by intuition). The science of phenomenology, then, is the antithesis of classical science: it is holistic rather than analytical; descriptive rather than explanatory; it emphasises the individual and unique rather than generalising; it values subjective insight and intuition rather than 'objectivity' and collapses the subject-object dualism; it sees man and nature as one and the same; and it is infused with notions of free will rather than determinsim.

PHYSICO-THEOLOGY Theology is the attempt to talk rationally about the divine; physico-theology is the attempt to explain the physical world in terms of the logic of God as the FINAL CAUSE of events. Physico-theology explanations hold that the universe is ruled by physical principles or laws which are a function of, and help to achieve, God's purpose or design for it. (Glacken 1967, see teleological).

PLENITUDE Fullness, completeness, plenty. The 'principle of plenitude' originates with Plato's *Timaeus* (Glacken). It observes that there is an abundance and rich variety of creatures and species on earth, and recognises a natural tendency for this to be so. Individuals and species tend to multiply freely so as to fill every niche and habitat which the earth provides.

PLURALISM A term used in politics for the situation whereby a society contains no particularly dominant interest groups - political,economic, ethnic, etc. Thus, decision making occurs via a process of democratic competition between many (a plurality of) groups, advocating and debating their particular point of view, and being heard by relatively impartial people (the judiciary, inspectors in public inquiries, government officers, the public at large). The outcome of such a process is often a compromise, whereby all groups have their interests accommodated to an extent, but not fully. This situation contrasts with one where the interests of an elite are dominant, and decisions are undemocratically made. In a pluralist society change is held to occur in response to 'stresses' articulated by particular groups. The stresses are accommodated to a degree by the rest of society through the democratic process. Sandbach (1980) draws the analogy between this social model and the systems model of the natural world.

POSITIVISM An approach to knowledge which involves attempting to make law-like statements that describe and explain observed (empirically-recognised) regularities in relationships between variables. This approach is basic to the method and philosophy of classical science, and to the philosophy of LOGICAL POSITIVISM (or logical empiricism). This term refers to how meaningful knowledge can be verified - first by breaking down, or analysing, the statements which comprise such knowledge, via *logic*, into 'elementary' statements; second, by testing the statements against observed reality. It follows that logical positivism holds as meaningless or unimportant any knowledge which cannot be verified (or falsified) via these procedures - that is, knowledge gained by intuition, emotional insights, religious or superstitious beliefs etc. Thus metaphysics are eliminated and positivist science claims to be 'value-free' (objective), and that value-free knowledge, founded in 'positive' 'facts', is superior to values-influenced (subjective) knowledge.

PROLETARIAT: see Capitalism.

REDUCTIONISM The breaking down (i.e. ANALYSIS) of complex things into simple constituents. The view (in opposition to *holism*) that a system or a phenomenon can be understood in terms of a knowledge of its isolated parts, rather than of how the parts relate to form a whole. Reductionism might involve conveying mental and spiritual processes solely in terms of physiological and physico-chemical events, or describing social relationships purely in terms of material economic quantities, or defining humans, animals and plants as simply physico-mechanical systems, or expressing philosophy merely as logical analysis.

ROMANTICISM A complex movement, amounting to a philosophy, which particularly affected Europe and N. America in the late-18th and early-19th centuries, but has legacies today, e.g. in ecocentrism. Some of its characteristics are: rejection of rationalism and the fruits of modern science and technology; desire to explore the realms of the subjective - e.g. of imagination and (deeply passionate) feeling; rejection of (urban) sophistication for (supposedly rural) simplicity and innocence; rejection of formal rules and restrictions of the individual in art and in ideas about society; a spiritual affinity for nature as purifying man's soul and as an expression of God; a liking for myth, fable and hallucination concerning heroic and fantastic events and deeds. The movement had many manifestations, and it is therefore of only limited use to generalise about it.

SENESCENCE (or natural senescence). The idea that the earth is ageing and deteriorating since its creation - possibly as a result of the sins of mankind. The effect of this on man and the environment would be adverse - producing population decline, soil infertility and erosion, moral decline etc. The idea arises from the analogy of earth with a living organism, and as Glacken (1967 p379) points out, it stands in the way of the 'essentially optimistic' idea that the order of nature on earth results from the beneficent design of the Creator. The idea was supplanted in the 18th century by those who argued for a constancy in nature and that population was not declining.

TECHNOCENTRISM A 'mode of thought' (O'Riordan) which recognises environmental 'problems' but believes either unrestrainedly that man will always solve them and achieve unlimited growth (the 'cornucopians') or, more cautiously, that by careful economic and environmental management they can be negotiated (the 'accommodators'). In either case considerable faith is placed in the ability and usefulness of classical science, technology, conventional economic reasoning (e.g. cost-benefit analysis), and their practitioners. There is little desire for genuine public participation in decision making, especially to the right of this ideology, or for debates about values. The technocentric's veil of optimism can be stripped away to reveal 'uncertainty, prevarication and tendency to error'. He envisages no radical alteration of social, economic or political structures, although those on the 'left' are gradualist reformers.

TELEOLOGICAL Embodying the idea of purpose or design. Teleological explanations in science are and were those describing events in terms of future and 'higher' goals rather than as a result of the causal influence of past events. Thus, medieval cosmology had objects falling to earth because of their striving to achieve their rightful place among the elements. The purpose here was God's, but it did not have to be; for example many modern biological explanations are teleological. Events are said to happen in a biological system *in order to* bring it back to equilibrium. Thus, while a physicist observes that raising a gas's temperature causes its volume to increase he does not say that this is in order to maintain thermal equilibrium; yet a physiologist, in observing that increasing a dog's temperature causes its respiration to increase does say that this is in order to oppose the temperature rise and keep the dog's system stable (Richards 1983) (see physico-theology).

UNIVERSALISE A term used frequently in Marxism for the presentation of a proposition or a set of circumstances or an ideology as being generally true in time and space and therefore obtaining in all circumstances (such as the 'law of supply and demand'). It is then devoid of historical context, i.e. is a-historical. This frequently serves to justify or legitimate the proposition etc. - for if it is universally true or 'natural', then it is inevitable and often therefore more acceptable.

INDEX

Agrarian socialism 193-4
Agriculture 165,171,189-90,194
Alienation 151-2,157-8,165-6,170,235; in education 220
Anarchism 190-3,200,202,235
A-political stances 174-7,201,203-4,208; of education 216
Assimilation of protest 183-4

Bacon, Francis 8,44,58,192,218; and the nature of science 54-5, 101,165; and the nature of scientists 55-7,62
Baconian Creed 55,60,106,116
Bahro, R. 203-4
Base-superstructure model 149-151,156,158,160,178,184; and education 217,222-4
Bioethic 27,71-2,79,82,88-9,235
Blueprint for Survival 24-6,113,156,188-93, *passim*, 198-200 *passim*,204,207,213,215,224
Bookchin, M. 14,109-10,183-4,191-2,202-3,205,236

Capitalism 149,153-4,156-7,164-6,167-73,236 *passim*,178, 185,201-3 *passim*; and education 219-23; and growth 170,176-7
Capra, F. 5,50,74,89,101,114,121-2,125,195,202
Carlyle, Thomas 77
Carrying capacity 23,91,99,111,209
Chase, A 95,98,132,208,210,212
Checks to population 94
Christianity and nature 44-6
City, as profane 76,84-6,87,166,198
Class 164,170,201,203,219; bourgeoisie and proletariat 150-3,156-7,235; environmentalism and class interests 174-6,178,180-1,183-4,208-13
Classical science 37-9,46-52,54-67
Club of Rome 24,143-4
Commoner, B. 20-1,202
Communism 152,191-3,197,209,213
Condorcet, Marquis de 57-9 *passim*,92
Consciousness; and education 216-7,219-20,223; and perception 125; in Marxist analysis 151,154,161; in phenomenology 119-21; in subatomic physics 122-4; social

consciousness 184-5,204,225
Cooperatives 195-6,200,224-5
Copernicus 46-7
Cotgrove, S. 21,28,32,127,144
Cultural filter 6-7,11,37,125,136,156
Curriculum centralisation 222-3

Darwin, Charles 100-03; social Darwinism 134-6,139,174,206-7,219
Decentralisation 188-96,200 (see also 'small communities')
Descartes, René 50-2,54; Cartesian dualism 50-2,79,117-19,122-4,163,236
Determinism; and education 217,221; and Marxism 153-6,158-62, 164,173,174; and the Great Chain of Being 74-5; environmental 34-5, 81,110-13,160,236; scientific 48,60,96,112,117-8,122-4,132,134-6, 139
Dialectic 148,155-6,161- 4,214, 223,236
Dominion over nature (man's) 44-6,55,60,106,110,116,192

Ecocentrism 27-8,31,33,43,Ch 3,Ch 4,111,180,237; and education Ch 8; and Marxism 173-7; and politics Ch 7; and romanticism 83,86-90
Ecofascism 194,204-13; elements of 204-6
Ecologist, The 14,191
Ecology Party 198-201 *passim*,207,213
Economic Laws 25-6,29,35,139,150,153,155,178; and education 221
Economic reductionism 21-2,66,149; and objectification 151-3, 160,165-6
Ecosocialism 188-96; elements of 197-201
Ecosystems perspective 28; and the Great Chain of Being 73-5, 101-110,237
Ecotopia 25,73,188,192,200,202,205-8,224
Education; and capitalism 219-23; and indoctrination 216, 219-20,223; and objectivity 218-9; and

243